P9-CBP-928

Mathematics
Teaching Today

Mathematics Teaching Today

Improving Practice, Improving Student Learning

Second Edition

Tami S. Martin, *Editor*

WRITING GROUP

Terese Herrera

Timothy D. Kanold

Roberta K. Koss

Patrick Ryan

William R. Speer

Originally published as *Professional Standards for Teaching Mathematics,* 1991

NATIONAL COUNCIL OF
TEACHERS OF MATHEMATICS

Library of Congress Cataloging-in-Publication Data

Mathematics teaching today : improving practice, improving student learning / Tami S. Martin, editor ; writing group, Terese Herrera ... [et al.]. — 2nd ed.

p. cm.

Includes bibliographical references.

ISBN 978-0-87353-598-4

1. Mathematics—Study and teaching—United States. I. Martin, Tami S. II. Herrera, Terese.

QA13.M165 2007

510.71'073—dc22

2007002253

The National Council of Teachers of Mathematics is a public voice of mathematics education, providing vision, leadership, and professional development to support teachers in ensuring mathematics learning of the highest quality for all students.

Printed in the United States of America

Credits

For the Second Edition

Professional Standards Revision Task Force Writing Group

Tami S. Martin, *Chair*	Illinois State University
Terese A. Herrera	The Ohio Resource Center for Mathematics, Science and Reading, Columbus, Ohio
Timothy D. Kanold	Adlai E. Stevenson High School, Lincolnshire, Illinois
Roberta K. Koss	T^3 (Teachers Teaching with Technology), San Rafael, California
Patrick Ryan	McGill University, Montreal, Quebec
William R. Speer	University of Nevada—Las Vegas
Harry B. Tunis, *Staff Liaison*	NCTM Headquarters Staff

For the Original Edition

The NCTM Commission on Teaching Standards for School Mathematics

Glenda Lappan, *Chair*	Michigan State University
Iris M. Carl	Houston Independent School District
Shirley Frye	Scottsdale School District
James D. Gates	NCTM Executive Director

Board Liaison

Lee V. Stiff	North Carolina State University

Working Group Members

Mathematics Teaching

Deborah Ball, *Chair*	Michigan State University
Evelyn Bell	Ysleta Independent School District, El Paso, Texas
Roberta Koss	Redwood High School, Larkspur, California
Steve Krulik	Temple University
Jane Schielack	Texas A&M University
Thomas Schroeder	University of British Columbia

Evaluation of Mathematics Teaching

Thomas Cooney, *Chair*	University of Georgia
Donald Chambers	Wisconsin State Department of Education
Marilyn Hala	NCTM Headquarters Staff
Tim Kanold	Stevenson High School, Prairie View, Illinois
Diane Thiessen	University of Northern Iowa
Sue Poole White	Banneker Senior High School, Washington, D.C.

Professional Development of Teachers of Mathematics

Susan Friel, *Chair*	University of North Carolina, Chapel Hill
Nicholas Branca	San Diego State University
Bettye Clark	Clark Atlanta University
Julie Keener	Hillside Junior High School, Boise, Idaho
James Leitzel	Ohio State University
Garry Musser	Oregon State University
William Speer	Bowling Green State University

Contents

Preface to the Second Edition

Mathematics Teaching Today: Improving Practice, Improving Student Learning, Second Edition updates the NCTM groundbreaking publication *Professional Standards for Teaching Mathematics,* published in 1991. Although much of the material in that landmark document remains sound and up to date, in spring 2003 the NCTM Board of Directors determined that recent developments in the field—including the Council's 2000 publication of *Principles and Standards for School Mathematics*—warranted this update of the original Professional Teaching Standards volume. The Board's decision was based on its concurrence with recommendations by several NCTM advisory committees that an update of the then twelve-year-old document was needed.

In summer 2003, the Professional Standards Revision Task Force, appointed by then NCTM President, Johnny Lott, convened in Bloomington, Illinois. The task force consisted of three authors who were part of the 1989 Commission on Professional Teaching Standards, whose members authored the original document, and three additional professional mathematics educators chosen for their expertise in each of the three major areas addressed in the document. Subsequent meetings of task force members were held in Las Vegas, Nevada; Anaheim, California; and Bloomington, Illinois.

The authors of *Mathematics Teaching Today* have strived to keep the focus and spirit of this updated volume consistent with those of the original document. Our aim is to articulate a vision for effective mathematics teaching as well as describe the support systems that are required to achieve that vision. We acknowledge the crucial roles that teachers, supervisors, principals, teacher-development professionals, parents, and communities all play in bringing the vision of high-quality mathematics teaching to life. In our quest for more and better mathematics for all, everyone has an important responsibility to our students—ensuring that all children have the opportunities they need to learn to their fullest potential.

Acknowledgments

Any document with the ambitious goal of describing the conditions that foster the best mathematics teaching requires the time, energy, and commitment of many talented individuals. The members of the Professional Standards Revision Task Force are indebted to many people for helping us takes steps toward reaching our goal. First, we thank the authors, editors, and others who contributed to the original Professional Teaching Standards document. The high-quality material already in place in that volume made our jobs all that much easier to perform.

Our work was encouraged and supported by many in the communities of professional organizations related to mathematics and the teaching and learning of mathematics. We are indebted to the individuals who eagerly answered the call to read and respond to drafts of our manuscript in late summer 2005. The comments of these reviewers helped us reflect on and incorporate perspectives and issues that may not have been given sufficient consideration in our initial drafts.

We are grateful for the consistent support of the NCTM Board of Directors; Presidents Johnny Lott, Cathy Seeley, and Francis (Skip) Fennell; as well as other leadership personnel at the Council. We would also like to especially thank our NCTM staff liaison, chief encourager, and consummate professional, Harry Tunis, whose calm demeanor and expert guidance were invaluable on this project. We gratefully acknowledge the tireless efforts of the NCTM editorial and production staff, especially Nancy Busse, Ann Butterfield, and Randy White, who improved the manuscript immensely and saw this project through to its completion.

We also thank Roger Day, Cynthia Langrall, and Corey Andreasen, who brought their expertise to the project when it was most needed, during the final rewrite. We acknowledge the help of Halcyon Foster, who took notes at one of our first planning meetings, freeing the rest of us to think and talk without having to focus on making a record of the dialogue.

Lastly, but always of primary importance, we are grateful to our families for their patience, love, and support. Volunteer activities such as this one always require the most sacrifice from the families of the volunteers. We appreciate their support of our mission to contribute in some small way to the mission of NCTM and the improvement of the teaching and learning of mathematics.

Tami S. Martin, *Editor,*
for the Writing Group

Preface to the Original Edition

In early 1989 the Commission on Professional Teaching Standards was established by the Board of Directors of the National Council of Teachers of Mathematics. The commission was charged to produce a set of standards that promotes a vision of mathematics teaching, evaluating mathematics teaching, the professional development of mathematics teachers, and responsibilities for professional development and support, all of which would contribute to the improvement of mathematics education as envisioned in the *Curriculum and Evaluation Standards for School Mathematics.*

The standards were drafted in the summer of 1989 and revised in the summer of 1990 by the members of the commission and the three working groups, each representing a cross section of the mathematics education community, including classroom teachers, supervisors, educational researchers, mathematics teacher educators, and university mathematicians. They were appointed by Shirley Frye, then president of NCTM.

The meetings of the commission and the working groups were held at Michigan State University. All who worked on the document thank Frank Hoppensteadt, Dean of the College of Natural Science, and Judith Lanier, Dean of the College of Education, and the many faculty members who helped to make working conditions for the group ideal. In addition we owe a debt of gratitude to Nan Jackson, Janine Remillard, and Kara Suzuka for their excellent contribution to the final drafting of the document and for their fine work in organizing and coding the hundreds of written responses to the first draft of the document.

The revisions were based on the very thoughtful responses to the working draft of this document gathered from individuals and groups during the 1989–90 school year. This document is much stronger and more coherent because of the careful reviews and suggestions that were provided. We thank all who contributed comments and hope that you see the results of your reactions in this final document. The stories of teaching that are used to craft the vignettes represent many of your suggestions. We are confident that this document represents the consensus of NCTM's members about teaching mathematics, evaluating the teaching of mathematics, the professional development of teachers, and responsibilities for the support and development of teachers and teaching.

The *Professional Standards for Teaching Mathematics* is designed, along with the *Curriculum and Evaluation Standards for School Mathematics,* to establish a broad framework to guide reform in school mathematics in the next decade. In particular, these standards present a vision of what teaching should entail to support the changes in curriculum set out in the *Curriculum and Evaluation Standards.* This document spells out what teachers need to know to teach toward new goals for mathematics education and how teaching should be evaluated for the purpose of improvement. We challenge all who have responsibility for any part of the support and development of mathematics teachers and teaching to use these standards as a basis for discussion and for making needed change so that we can reach our goal of a quality mathematics education for every child.

Acknowledgments from the Original Edition

Preparation for the NCTM standards originated more than a decade ago with the release in 1980 of *An Agenda for Action: Recommendations for School Mathematics [of] the 1980s.* Since then, the goals and activities of the Council have sustained and shaped the emphasis on curricular reform. The continued commitment to the evolving process of change in school mathematics has become an NCTM baton, passed on by each president to the next. This document, the companion to the *Curriculum and Evaluation Standards for School Mathematics,* represents the current step in the continuum: it advances and expands the vision of a high-quality mathematics education for every child.

The Council acknowledges with gratitude the outstanding leadership of Glenda Lappan, who chaired the Commission on Teaching Standards for School Mathematics and effectively directed the writing project. The writers and working groups deserve great credit for their scholarship, diligence, and dedication. We are also grateful for the active participation of NCTM's membership in the review process along with that of the other organizations and thousands of individuals both within and outside the profession who volunteered their reactions.

We appreciate the support and total involvement of the Headquarters staff throughout every stage of the project. Our executive director, James Gates, who guided that vital process, deserves full recognition and our thanks.

F. Joe Crosswhite, President 1984–1986
John A. Dossey, President 1986–1988
Shirley M. Frye, President 1988–1990
Iris M. Carl, President 1990–1992

Introduction

Background and Rationale

Calls for reform in mathematics education are prompted by several factors. Large-scale international and national assessments have generated ample evidence that, on average, United States and Canadian students' mathematical knowledge and skills are lacking (Mullis et al. 2000; Mullis, Martin, Gonzalez, and Chrostowski 2004; National Commission on Mathematics and Science Teaching for the 21st Century 2000). That situation persists despite a growing need for a mathematically literate workforce in an increasingly technology-based global community (NCTM 2000). Because performance in mathematics has been tied to success in professional careers, particularly those that lead to economic and political advantages, interest in mathematics education has spread far beyond school walls. Calls for reform, however, are just that—a way to alert a community to an existing deficiency and, perhaps, offer suggestions for how to remedy the ills that have led to the deficiency.

This book delineates Standards for various aspects of the teaching profession, including teachers' practice, professional supervision, collegial interaction, and career-long professional growth. Together, these Standards provide a framework for professional practice that supports the NCTM's vision of more and better mathematics for all children. In addition, this book explicates the roles of teachers, supervisors, teacher educators, mathematicians, professional developers, parents, politicians, community members, and others in improving the teaching and learning of mathematics.

A Vision of High-Quality Teaching and Learning

In April 2000, the National Council of Teachers of Mathematics (NCTM) released its *Principles and Standards for School Mathematics.* Those Principles and Standards represent a vision of school mathematics that has been continually developed and refined since the release of the landmark trio of *Standards* documents—*Curriculum and Evaluation Standards for School Mathematics* (1989), *Professional Standards for Teaching Mathematics* (1991), and *Assessment Standards for School Mathematics* (1995). The initial three documents, along with *Principles and Standards for School Mathematics* (2000) and the many subsequent text and online support resources (e.g., the Addenda series, the Navigations series, and the Illuminations project), represent a historically significant effort to articulate goals for the improvement of mathematics teaching and learning.

The collection of NCTM *Standards* documents and resources represents years of planning, writing, and consensus-building among the membership of NCTM and the broader mathematics, science, engineering, and education communities; the business community; parents; and school administrators. The *Standards* documents collectively describe elements that contribute to high-quality mathematics education for United States and Canadian students in prekindergarten through grade 12. The vision for school mathematics outlined in *Principles and Standards for School Mathematics* embodies the collective goals of the Standards movement to date:

The Standards documents collectively describe elements that contribute to high-quality mathematics education for North American students in prekindergarten through grade 12.

> Imagine a classroom, a school, or a school district where all students have access to high-quality, engaging mathematics instruction. There are ambitious expectations for all, with accommodation for those who need it. Knowledgeable teachers have adequate resources to support their work and are continually growing as professionals. The curriculum is mathematically rich, offering students opportunities to learn important mathematical concepts and procedures with understanding. Technology is an essential component of the environment. Students confidently engage in complex mathematical tasks chosen carefully by teachers. They draw on knowledge from a wide variety of mathematical topics, sometimes approaching the same problem from different mathematical perspectives or representing the mathematics in different ways until they find methods that enable them to make progress. Teachers help students make, refine, and explore conjectures on the basis of evidence and use a variety of reasoning and proof techniques to confirm or disprove those conjectures. Students are flexible and resourceful problem solvers. Alone or in groups and with access to technology, they work productively and reflectively, with the skilled guidance of their teachers. Orally and in writing, students communicate their ideas and results effectively. They value mathematics and engage actively in learning it.
>
> (NCTM 2000, p. 3)

To make progress toward that vision, *Principles and Standards* outlines six Principles and ten Standards to guide decision making about and within classrooms. The Principles reflect precepts that are fundamental to a high-quality mathematics education. Those Principles include Equity, Curriculum, Teaching, Learning, Assessment, and Technology. Those six Principles, although not unique to mathematics, influence decisions related to curriculum, pedagogy, and teacher development and support. As a result, they form the foundation for more specific recommendations related to the teaching and learning of mathematics.

The Standards also describe aspects of school mathematics that are valued. Compared with the Principles, the Standards are more specific statements about what students should know and be able to do. The five Content Standards—Number and Operations, Algebra, Geometry, Measurement, and Data Analysis and Probability—summarize the content goals for students in prekindergarten through grade 12. The five Process Standards—Problem Solving, Reasoning and Proof, Connections, Communications, and Representations—describe ways of acquiring and using content knowledge.

NCTM is not alone in identifying the importance of both content and processes in school mathematics. The National Research Council (NRC) defines *mathematical proficiency* as being composed of five intertwined strands (NRC 2001):

1. Conceptual understanding—comprehension of mathematical concepts, operations, and relations
2. Procedural fluency—skill in carrying out procedures flexibly, accurately, efficiently, and appropriately
3. Strategic competence—ability to formulate, represent, and solve mathematical problems

4. Adaptive reasoning—capacity for logical thought, reflection, explanation, and justification
5. Productive disposition—habitual inclination to see mathematics as sensible, useful, and worthwhile, coupled with a belief in diligence and one's own efficacy

The parallels between the mathematical proficiency called for by the NRC and the rich mathematical experiences called for in *Principles and Standards for School Mathematics* are many, including incorporation of the mathematical processes of problem solving, reasoning, communications, connections, and representation. Perhaps the most important of those parallels, however, is the stated belief that "all students can and should be proficient in mathematics" (NRC 2002, p. 1).

Justification for Teaching Standards

One of the most important reasons to create and subsequently to update the original *Professional Standards for Teaching Mathematics* (NCTM 1991) was to underscore the following message: More than curriculum standards documents are needed to improve student learning and achievement. Teaching matters. Therefore, exploring what goes on in mathematics classrooms is essential to identifying issues and looking for opportunities for improvement.

More than curriculum standards documents are needed to improve student learning and achievement. Teaching matters.

One predominant method of instruction in U.S. schools has prevailed for almost a century (Hiebert 1999; National Commission on Mathematics and Science Teaching for the 21st Century 2000; Stigler and Hiebert 1999). In typical U.S. mathematics classrooms, teachers first spend a considerable portion of class time checking homework, demonstrating procedures, and asking and answering questions related to those procedures. Next, students engage in seatwork to practice procedures. Finally, class closes with the assignment of homework for the next day.

That pattern of classroom interactions was well documented by the descriptions of eighth-grade mathematics classes drawn from the 1995 video study associated with the Third International Mathematics and Science Study (TIMSS). In that study, typical teaching in the United States was characterized as "learning terms and practicing procedures" (Stigler and Hiebert 1999, p. 27). The researchers found that the mathematical content in U.S. eighth-grade lessons was about a year behind the content being taught in Germany and Japan, and that much less mathematical reasoning was required of U.S. students. U.S. teachers typically presented definitions and demonstrated procedures for solving specific problems; their students were responsible for memorizing the definitions and practicing the procedures.

Interviews in the 1995 study also led researchers to conclude that most U.S. teachers believed they were implementing standards-based teaching methods that emphasize higher-order thinking. This evidence showed that many U.S. teachers have made good-faith efforts to incorporate recommendations made in NCTM *Standards* documents. However, misinterpretations of reform recommendations or implementation of superficial changes, such as more group work, calculators, manipulatives, or real-world problems, may account for the apparent contradictions between teachers' self reports and researchers' observations in the video study (Stigler and Hiebert 1999).

The follow-up 1999 TIMSS video study further showed that most of the problems that U.S. eighth-grade students solve in class are at a lower level of complexity than the problems solved by their counterparts in higher-performing countries. Similarly, U.S. students spend the majority of their time solving problems that focus on using procedures as compared with students in higher-performing countries, who spend the majority of their time solving problems that focus on making connections.

Overall, however, the 1999 video study demonstrated that no single approach to teaching mathematics is used by teachers in higher-achieving countries (Hiebert et al. 2003a; 2003b). Although signature patterns of instruction were found in each country, variability was also found in teaching strategies within countries. The study's authors concluded that questions about teaching have no simple answers. Rather, teaching is a complex endeavor, and much more remains to be learned about connections between teaching and learning.

The recommendations in this volume were guided by literature on teaching, professional development, and professional practice.

Why *should* teachers follow the recommendations made in this and other NCTM Standards documents? Have these Standards been proved effective by educational research? Unlike scientific research that is conducted in a laboratory in which all factors can be identified or controlled, schools have too many variables that cannot be identified, let alone controlled in a truly randomized fashion. Likewise, treatments cannot be randomly assigned or consistently applied (Hiebert 1999, 2003) However, an extensive research base on children's thinking and learning was used to guide the assumptions and recommendations made in *Principles and Standards for School Mathematics* (Kilpatrick, Martin, and Schifter 2003). Similarly, the recommendations in this volume were guided by literature on teaching, professional development, and professional practice.

For example, evidence has been mounting in the literature that use of standards-based curriculum materials by teachers who implement them using standards-based teaching practices has a positive affect on student achievement (Fuson, Carroll, and Drueck 2000; McGaffrey, Hamilton, Stecher, Klein, Bugliari, and Robyn 2001; Reys, Reys, Lapan, Holliday, and Wasman 2003; Riordan and Noyce 2001; Schoen, Cebulla, Finn, and Fi 2003). In particular, the use of NCTM Standards–aligned curriculum materials does not have a detrimental effect on students' abilities to perform calculations but often has a positive effect on students' ability to perform tasks that require higher-order thinking processes, such as problem solving and reasoning (Fuson, Carroll, and Drueck 2000; Lappan and Bouck 1998). Perhaps not surprisingly, students who had opportunities to engage in reasoning and problem solving were better able to perform those processes. In other words, students learned what they had opportunities to learn (Hiebert 1999). For that reason, the choices teachers make about what to teach and how to teach it are crucial.

The National Commission on Mathematics and Science Teaching for the 21st Century (2000) came to a similar conclusion regarding the important roles played by the teacher. In the foreword to what has become known as the Glenn Commission Report, *Before It's Too Late: A Report to the Nation from the National Commission on Mathematics and Science Teaching for the 21st Century*, the commission acknowledged the vital role of mathematics and science teaching (2000, p. 5):

> First, at the daybreak of this new century and millennium, the Commission is convinced that the future well-being of our nation and people depends not just on how well we educate our children generally, but on how well we educate them in mathematics and science specifically.... Second, it is abundantly clear from the evidence already at hand that we are not doing the job that we should do—or can do—in teaching our children to understand and use ideas from these fields. Our children are falling behind; they are simply not "world-class learners" when it comes to mathematics and science.... Third, after an extensive, in-depth review of what is happening in our classrooms, the Commission has concluded that the most powerful instrument for change, and therefore the place to begin, lies at the very core of education—*with teaching itself.*

Teaching is a deeply entrenched cultural phenomenon. Behaviors and patterns of interaction occur as a result of beliefs and expectations that may exist at a subconscious level and can be changed only slowly over time, with great effort, self-reflection, and commitment (Stigler and Hiebert 1999). In addition, teachers cannot make those changes without accompanying changes in the support system that comprises much of the teaching and school culture. Supervisors, principals, parents, and other concerned adults must provide financial, structural, and moral support to help teachers truly focus on the mathematical understanding of all students (NRC 2001).

Supervisors, principals, parents, and other concerned adults must provide financial, structural, and moral support to help teachers truly focus on the mathematical understanding of all students.

Roles of Teachers and Others: A Shift in Perspective

To reach the ambitious goals outlined in *Principles and Standards for School Mathematics* and other Standards documents, classroom environments, teaching practices, teacher development programs, school- and district-level administrators, parents, and community members must support the vision outlined in those documents. Teachers, specifically, must shift their perspectives about teaching, from that of a process of delivering information to that of a process of facilitating students' sense making about mathematics. That shift will require teachers in prekindergarten through grade 12 to be proficient in—

- designing and implementing mathematical experiences that stimulate student's interests and intellect;
- orchestrating classroom discourse in ways that promote the exploration and growth of mathematical ideas;
- using, and helping students use, technology and other tools to pursue mathematical investigations;

- assessing students' existing mathematical knowledge and challenging students to extend that knowledge;
- fostering positive attitudes about the aesthetic and utilitarian values of mathematics;
- engaging in opportunities to deepen their own understanding of the mathematics being studied and its applications;
- reflecting on the value of classroom encounters and taking action to improve their practice; and
- fostering professional and collegial relationships to enhance their own teaching performance.

This book, along with others in the Standards collection, is a part of the NCTM's commitment to championing the cause of improvement in the teaching and learning of mathematics. The NCTM Standards books, supplements, and resources can be used as frameworks and exemplars to guide the work of teachers, administrators, school districts, states, provinces, certification boards, university faculty, and other groups in their quests to propose solutions to curricular, instructional, assessment, supervision, and teacher development issues in mathematics education in their own communities.

Assumptions about Teaching and Learning

Mathematics Teaching Today rests on the following two assumptions:

- Teachers are essential figures in improving the ways in which mathematics is taught and learned in schools.
- Such improvements require that teachers have long-term support and adequate resources.

Good instructional and assessment materials and the latitude to use them flexibly are also important elements of change.

The kind of teaching envisioned in these Standards is significantly different from what many teachers themselves have experienced as students in mathematics classes. Because teachers need time to learn and develop Standards-based teaching practice, appropriate and ongoing professional development is crucial. Good instructional and assessment materials and the latitude to use them flexibly are also important elements of change.

For teachers to be able to change their role and the nature of their classroom environment, administrators, supervisors, and parents must expect, encourage, support, and reward the kind of teaching described in this set of Standards. We cannot expect teachers to respond simultaneously to several different calls for change or other new demands. Change is difficult and requires time and reliable, systematic support.

Major Shifts in Classroom Environment

Woven into the fabric of *Mathematics Teaching Today* are six major objectives in the environment of mathematics classrooms that we need to move toward:

- Communities that offer an equal opportunity to learn to all students
- A balanced focus on conceptual understanding as well as on procedural fluency
- Active student engagement in problem solving, reasoning, communicating, making connections, and using multiple representations
- Technologically well-equipped learning centers in which technology is used to enhance understanding
- Incorporation of multiple assessments that are aligned with instructional goals and practices
- Mathematical authority that lies within the power of sound reasoning and mathematical integrity

To achieve that vision, schools and communities must support the work of teachers and must provide the time needed for professional collaboration, planning, and enhancement of teachers' knowledge of mathematics and pedagogy.

Schools and communities must support the work of teachers and must provide the time needed for professional collaboration, planning, and enhancement of teachers' knowledge of mathematics and pedagogy.

Equity

The theme of equity is addressed throughout these Standards. Equity is defined as "high expectations and strong support for all students" (NCTM 2000, p. 11). More specifically, equity requires "high expectations and worthwhile opportunities for all, ... accommodating differences to help everyone learn, ... [as well as] resources and support for all classrooms and all students" (NCTM 2000, pp. 12–14). By using the phrase *all students,* we mean to set the mathematical education of every child as the goal for mathematics teaching at all levels, prekindergarten through grade 12.

By "all students" we mean specifically—

- students who have been denied access in any way to educational opportunities as well as those who have not;
- students who are African Americans, Hispanics/Latinos, Native Americans, Alaskan Natives, Pacific Islanders, Asian Americans, First Nations people, and other minorities, as well as those who are considered to be a part of the majority;
- students who are female as well as those who are male;
- students who are from any socioeconomic background;

- students who are native English speakers and those who are not native English speakers;
- students with disabilities and those without disabilities; and
- students who have not been successful in school and in mathematics as well as those who have been successful.

Schools and communities must accept as essential the goal of mathematical education for every child.

Schools and communities must accept as essential the goal of mathematical education for every child. However, that requirement does not mean that every child will have the same interests or capabilities in mathematics. It does mean that we must examine our fundamental expectations about what children can learn and can do and that we must strive to create learning environments in which raised expectations for children can be met.

The Mathematics Teaching Cycle

To realize the vision for teaching and learning outlined in the collection of NCTM Standards documents, teachers must have extensive knowledge of mathematics, of learners, and of mathematical pedagogy as well the skills to implement plans and analyze the outcomes of practice. That collection of knowledge and skills has been characterized in many ways and can be viewed as a cycle of teaching activity (Reynolds 1992; Shulman 1987), as seen in figure I.1.

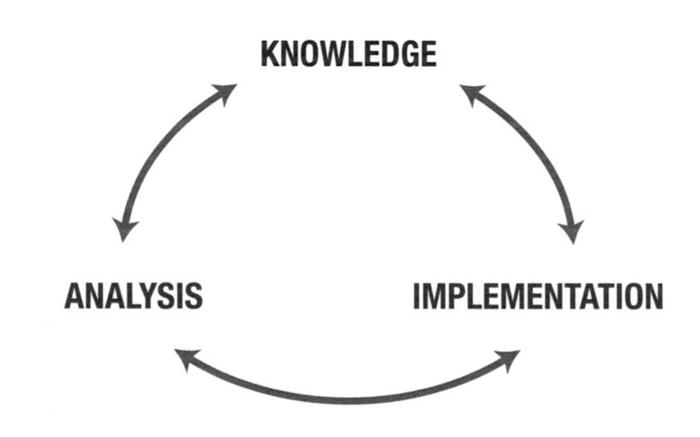

Fig. I.1. The cycle of teaching activity

The ways in which teachers are treated by their colleagues, their supervisors, school administrators, and others have an influence on how teachers perceive themselves as professionals and the goals they set for improving practice.

The cycle of teaching does not happen in a vacuum. Knowledge is developed in preparation for teaching, as a result of teaching experiences, and through career-long professional development. Similarly, the cultural environment of the school, district, community, and the profession influences the knowledge teachers bring to the classroom, the ways in which they implement classroom instruction and assessment, and the lenses through which they view classroom occurrences and their own performances as teachers. The ways in which teachers are treated by their colleagues, their supervisors, school administrators, and others have an influence on how teachers perceive themselves as professionals and the goals they set for improving practice. See figure I.2.

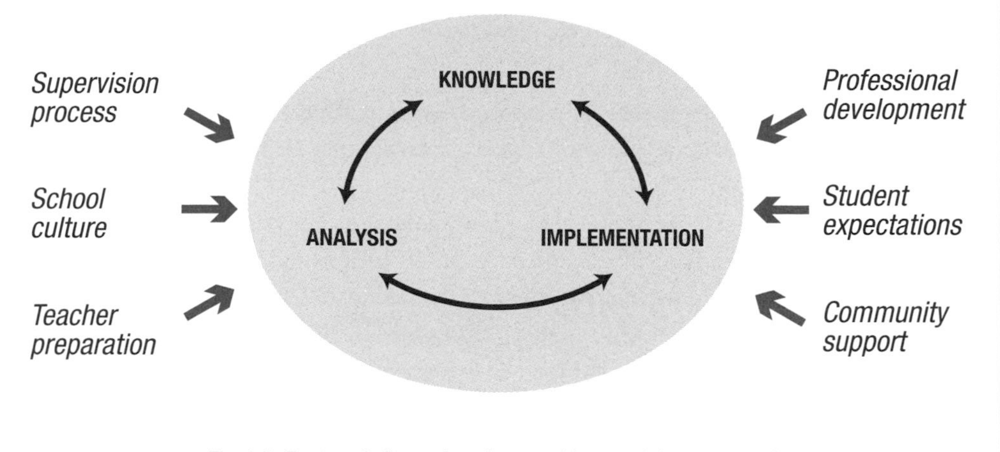

Fig. I.2. Factors influencing the teaching-activity cycle and teachers' self-perceptions

Components of *Mathematics Teaching Today*

Because the professional life of a mathematics teacher is multifaceted, we have developed these Standards to address several fundamental aspects of the mathematics teaching profession. In particular, this document contains four chapters that address the practice, support, and development of mathematics teachers:

Chapter 1: Standards for Teaching and Learning Mathematics

Chapter 2: Standards for the Observation, Supervision, and Improvement of Mathematics Teaching

Chapter 3: Standards for the Education and Continued Professional Growth of Teachers of Mathematics

Chapter 4: Working Together to Achieve the Vision

Standards for Teaching and Learning Mathematics

This section develops a vision of what a teacher at any level of schooling must know and be able to do to teach mathematics as described in the NCTM Standards documents, including *Principles and Standards for School Mathematics* (2000) and *Professional Standards for Teaching Mathematics* (1991). The Standards in this section are organized in a framework of teaching processes:

- Knowing about mathematics, general pedagogy, and students' mathematical learning, including modes of instruction and assessment, ways of building students' informal understanding, and ways to stimulate engagement and exploration
- Implementing teaching and learning activities, including selecting worthwhile mathematical tasks, creating a challenging and nurturing classroom

environment, and facilitating meaningful discourse that leads to socially negotiated understandings

- Analyzing students' learning, the mathematical tasks, the environment, and classroom discourse to make ongoing instructional decisions

The statements of Standards focus on major aspects of effective mathematics teaching across all grade levels. The elaboration sections further explain the intent and details of the Standards. The vignettes are used to illustrate various aspects of the Standards. Although the vignettes depict a range of situations in which mathematics teaching and learning can take place, limitations of space as well as the one-dimension medium of written communication do not allow for a comprehensive illustration of each Standard.

Taken together, the elaborations and vignettes offer a perspective on what constitutes exemplary teaching. In particular, a high-quality mathematics experience is not determined simply by the presence of the latest technological tools or the use of small groups, manipulatives, or student discussions. The nature of the mathematical task posed and what is expected of students are crucial aspects against which to judge the effectiveness of the lesson.

Standards for the Observation, Supervision, and Improvement of Mathematics Teaching

This section presents NCTM's vision for the assessment of teaching as well as the support required for ongoing improvement. The Standards address the process as well as the emphases of teacher observation, supervision, and improvement of teaching.

That cycle should incorporate a variety of data sources to inform participants in their analyses of teaching and learning.

The process that is described in this collection of Standards encompasses both the supervisor's and the teacher's roles in observation, supervision, and improvement. In particular, each is expected to contribute in a meaningful and supportive way to the continuous improvement cycle that characterizes the growth and development of the teacher. That cycle should incorporate a variety of data sources to inform participants in their analyses of teaching and learning. As a result, teachers and supervisors should develop mutually agreed on plans for professional growth that focuses on enhancing the teacher's ability to facilitate students' sense making of mathematics.

The focal points in the supervision process that are identified as Standards are primary indicators of quality in mathematics teaching. The extent to which teachers are creating successful learning experiences can be gleaned by attending to both what the teacher is doing and what the students are doing in the classroom. Through the lenses of teacher and student actions, supervisors can assess a teacher's ability to (a) select and implement worthwhile mathematical tasks, (b) orchestrate meaningful mathematical discourse within a supportive and engaging classroom environment, and (c) use multiple sources of information to analyze students' understanding.

These Standards give guidance to teachers seeking self-improvement, to colleagues mentoring others, and to supervisors and others who are involved in the observation, supervision, and improvement of teaching. The vignettes in this section depict a range of assessment activities and personnel involved in the processes associated with teacher assessment. They illustrate the nature of the process and some of the substantive points of discussion between supervisors and teachers and among teachers who work together to improve their instructional and assessment practices.

Standards for the Education and Continued Professional Growth of Teachers of Mathematics

This section expresses NCTM's vision for well-prepared teachers of mathematics from the time prospective teachers of mathematics take their first courses in collegiate mathematics throughout their career-long growth as teachers. These Standards focus on what a teacher needs to know about mathematics, mathematics education, and pedagogy to be able to carry out the vision of teaching discussed in this document. The following aspects of both the preservice and in-service phases of the professional growth of teachers are addressed:

- Modeling good mathematics teaching
- Knowing mathematics and school mathematics
- Knowing students as learners of mathematics
- Knowing mathematical pedagogy
- Developing as a teacher of mathematics
- Teachers' roles in professional growth

The Standards in this section provide essential guidance to colleges, universities, and schools; state departments and provincial ministries of education; public and private schools; and all who have a part in the preparation and professional growth of teachers. These Standards focus attention on the roles of faculty in college and university departments of education and mathematics and of school officials responsible for professional development. They also emphasize the need for dialogue among those partners in nurturing excellence in mathematics teaching.

The current reform movement in mathematics education, and in education in general, has the professionalism of teaching as a fundamental underlying theme. That view recognizes the teacher as a part of a learning community that continually fosters growth in knowledge, stature, and responsibility. The Standards in this section guide the preparation, support, and career development of teachers. The Standards themselves are meant to be general principles that can be used to improve the quality of teacher development programs as well as school and university professional development activities. The elaborations and vignettes illustrate applications of those general principles to levels of preparation and phases of career development.

Working Together to Achieve the Vision

This section spells out the responsibilities of those who make decisions that affect the teaching of mathematics. The responsibilities of the following groups are addressed:

- Policymakers in government, business, and industry
- Schools and school systems
- Colleges and universities
- Professional organizations

The environment in which teachers teach is as important to their success as the environment in which students learn is to theirs.

Decisions made by others may enable teachers to move toward the vision of teaching described in these Standards or may constrain the mathematics program in ways that cripple efforts to improve teaching. The environment in which teachers teach is as important to their success as the environment in which students learn is to theirs. This section highlights responsibilities of the groups listed above and describes ways in which others can support teachers in attaining the vision of teaching needed to implement *Principles and Standards for School Mathematics* (NCTM 2000).

In addition to the four chapters described in the foregoing, this volume includes a supplemental section titled "Questions for the Reflective Practitioner," which gives readers an opportunity to make connections between the document's recommendations and his or her own professional practice. Those questions were designed to inspire thoughtful reflection and discussion by individuals or within groups of preservice teachers, practicing teachers, school teacher leaders, department chairs, principals, district-level administrators, teacher educators, professional developers, parents, community members, and others with an interest in, and commitment to, improving the teaching and learning of school mathematics.

The book closes with a bibliography of references cited throughout the document as well as additional resources for further information. This list of sources is not meant to be comprehensive or compete, but rather is intended to give the reader a sense of some of the research and materials that are available to support the vision described in this book.

Conclusion

These Teaching Standards are not intended to be an exhaustive checklist of specific concepts, skills, and behaviors to be possessed by teachers. Rather, they are a set of principles accompanied by illustrations and indicators that can be used to judge what is valuable and appropriate. They give direction for moving toward excellence in teaching mathematics. They furnish guidance to all who are interested in improving teaching, including teachers, universities, state departments of education and provincial ministries of education, local school districts, private schools, teacher organizations, the National Council for Accreditation of Teacher Education (NCATE), the Teacher

Education Accreditation Council (TEAC), the National Board on Professional Teaching Standards, and others who license or certify teachers or who evaluate teaching or teacher education programs.

These Standards offer a vision for high-quality mathematics teaching and support structures for teachers. What matters in the long run, however, is how classrooms across the United States and Canada are transformed and how teaching practices evolve to address the mathematical learning needs of children.

Chapter 1

Standards for Teaching and Learning Mathematics

Overview

This section presents seven Standards for the teaching of mathematics organized under three categories: knowledge, implementation, and analysis.

Knowledge

1. Knowledge of Mathematics and General Pedagogy
2. Knowledge of Student Mathematical Learning

Implementation

3. Worthwhile Mathematical Tasks
4. Learning Environment
5. Discourse

Analysis

6. Reflection on Student Learning
7. Reflection on Teaching Practice

Introduction

Principles and Standards for School Mathematics represents NCTM's vision of school mathematics. That vision is designed to guide educators and others involved in the support of teaching and learning as they strive to improve classrooms and schools. The Principles of Equity, Curriculum, Teaching, Learning, Assessment, and Technology are the cornerstones of high-quality mathematics education. Those Principles describe high expectations and support for all students; a coherent, focused curriculum; teaching based on knowledge of students and content; meaningful learning that is actively constructed; multiple forms and purposes of assessment; and the appropriate use of technology to enhance understanding (NCTM 2000).

The vision for school mathematics that is articulated in *Principles and Standards* requires changes in what mathematics is taught and how it is taught. Teachers and students have

Teachers and students have different roles and different notions about what it means to know and to do mathematics.

different roles and different notions about what it means to know and to do mathematics. Teachers expect students to encounter, develop, and use mathematical ideas and skills in the context of genuine problems and situations. In so doing, students develop the ability to use a variety of resources and tools, such as calculators and computers as well as concrete, pictorial, and metaphorical models. They know and are able to choose appropriate methods of computation, including estimation, mental calculation, and the use of technology. As they explore and solve problems, they engage in conjecture and argument.

The purpose of this chapter is to sharpen and expand the images of teaching and learning mathematics to elaborate the vision set forth in *Principles and Standards for School Mathematics.*

Seven Standards represent the core dimensions of teaching and learning mathematics. Those Standards are organized under three headings—knowledge, implementation, and analysis—that represent major arenas of teachers' work that are central to defining what occurs in mathematics classrooms.

Effective mathematics teaching depends on a deep knowledge of mathematics. Teachers need to understand the big ideas of mathematics and be able to represent mathematics as a coherent and connected enterprise.

- *Knowledge* of teaching, of mathematics, and of students is an essential aspect of what a teacher needs to know to be successful. Effective mathematics teaching depends on a deep knowledge of mathematics. Teachers need to understand the big ideas of mathematics and be able to represent mathematics as a coherent and connected enterprise (Ball, Lubienski, and Mewborn 2001; Hill, Rowan, and Ball 2005; Ma 1999; Schifter, Russell, and Bastable 1999). In addition, teachers must have sound pedagogical knowledge and a grounded understanding of students as learners. Teachers use that knowledge to create learning communities that enable students to build a conceptual understanding of, and procedural proficiency with, mathematics.
- *Implementation* of learning activities within the classroom requires a teacher to choose worthwhile mathematical tasks, to establish a supportive and challenging environment, and to promote mathematical discourse among all members of the learning community. The learning environment results from the unique interplay of intellectual, social, and physical characteristics that shape the ways of knowing and working that are encouraged and expected in the classroom. Worthwhile mathematical tasks challenge students to make sense of both the contexts and the mathematics embedded in the tasks. The discourse of the learning community refers to the ways of representing, thinking, talking, and agreeing and disagreeing that teachers and students use as they engage in mathematical thinking and learning. Teachers, through the ways in which they orchestrate discourse, convey messages about whose knowledge and whose ways of thinking and knowing are valued, who is considered able to contribute, and who has status in the group. Discourse also reflects classroom values, such as what makes an answer right, what counts as legitimate mathematical activity, and what the standards are for an argument to be considered convincing.

- *Analysis* refers to the systematic reflection in which teachers engage. It entails the ongoing monitoring of classroom life: How well are the tasks, discourse, and environment fostering the development of every student's mathematical proficiency and understanding? Such systematic reflection may also involve colleagues' working together to analyze and improve teaching practice. Further, teachers use information garnered from both formative and summative assessment to guide instructional decisions. Through that process, teachers examine relationships between what they and their students are doing and what students are learning.

In deciding how to present and elaborate the ideas underlying each of the seven Standards, we confronted two basic challenges. First, teaching is an integrated activity. Although we have described elements of teaching as falling into three categories—knowledge, implementation, and analysis—they are, in fact, interwoven and interdependent. The quality of the classroom discourse, for example, is both a function of, and an influence on, the teacher's knowledge of mathematics. Similarly, mathematical tasks are shaped by the teacher's knowledge of her students and her reflections on the use of those tasks in other situations. Our second challenge was that professional standards for mathematics teaching should represent values about what contributes to good practice without prescribing it. Such standards should offer a vision, not a recipe.

The format of this chapter grew out of consideration of those challenges. Because teaching is an integrated activity and because we wanted to provide concrete images of a vision, we have chosen to use illustrative vignettes of classroom teaching and learning. The statement of each of the seven Standards is first elaborated with an explanation of its main ideas. Each explanation is then followed with one or more illuminating scenarios drawn from transcripts, observations, and experiences in a wide variety of real classrooms. The vignettes were selected to reflect a range of teaching styles, classroom contexts, mathematical topics, and grade levels. The vignettes were gathered from classrooms with students of diverse cultural, linguistic, and socioeconomic backgrounds and include examples of teachers facing problems as well as instances of accomplished practice.

Assumptions

The seven Standards for teaching are based on four assumptions about the practice of mathematics teaching.

1. *Principles and Standards for School Mathematics* furnishes the basis for a curriculum in which problem solving, reasoning and proof, communication, connections, and representation are central. Teachers must help all students develop conceptual and procedural understandings of number and operations, algebra, geometry, measurement, and data analysis and probability. They must engage all students in formulating and solving a wide variety of problems, making conjectures and constructing arguments, validating solutions,

and evaluating the reasonableness of mathematical claims. Teachers should make use of appropriate technologies to assist students in such endeavors. Teachers must also foster the disposition to use and engage in mathematics, an appreciation of its beauty and utility, and a tolerance for getting stuck or sidetracked. Teachers must help students realize that mathematical thinking may involve dead ends and detours, all the while encouraging them to persevere when confronted with a puzzling problem and to develop the self-confidence and interest to do so.

What students learn is fundamentally connected with _how_ they learn it.

2. *What* students learn is fundamentally connected with *how* they learn it. Students' opportunities to learn mathematics emerge from the setting and the kinds of tasks and discourse in which they participate. What students learn about particular concepts and procedures as well as about mathematical thinking depends on the ways in which they engage in mathematical activity with and without technology in their classrooms. Their disposition toward mathematics is also shaped by such experiences. Consequently, the goal of creating high-quality learning experiences for all students requires careful attention to pedagogy as well as to curriculum.
3. All students can learn to think mathematically. The vision described in *Principles and Standards for School Mathematics,* and as articulated in the Equity Principle, is a vision that applies to all students. Although not all students learn in the same ways, all students should be expected to engage in mathematical processes to develop the tools to make sense of mathematical ideas. More specifically, the activities of making conjectures, arguing about mathematics using mathematical evidence, formulating and solving problems—even perplexing ones—are not just for some group of students thought to be "bright" or "mathematically able" (Trentacosta and Kenney 1997). Every student can and should learn to reason and solve problems, to make connections across a rich web of topics and experiences, and to represent and communicate mathematical ideas with and without technology. In fact, engaging in those activities, or thinking mathematically, is the essence of doing mathematics.
4. Teaching is a complex practice. Teaching is composed of elements that interact with and reinforce one another (Stigler and Hiebert 1999). In particular, teaching mathematics draws on knowledge from several domains: knowledge of mathematics, of diverse learners, of how students learn mathematics, and of the contexts of classroom, school, and society. Such knowledge is general but not superficial. However, teachers must also take into account that teaching is context-specific. Theoretical knowledge about adolescent development, for instance, can only, in part, influence a decision about particular students learning a particular mathematical concept in a particular context. Teachers often find themselves balancing multiple goals and considerations as they weave together knowledge to decide how to respond to a student's question, how to represent a given mathematical idea, how long to pursue the discussion of a problem, or how to make appropriate use of available technologies to develop the richness of an investigation. Making appropriate decisions depends on a variety of factors that cannot be determined in the abstract or be governed by rules of thumb.

The challenge of teaching mathematics well depends on a host of considerations and understandings. Good teaching demands that teachers thoughtfully apply best available knowledge about mathematics, learning, and teaching to the particular contexts of their work. The Standards for teaching mathematics are designed to help guide the processes of such reasoning, highlighting issues that are crucial in creating the kind of teaching practice that supports the goals of *Principles and Standards for School Mathematics*. This chapter promulgates themes and values but does not—indeed, could not—prescribe "right" practice.

Knowledge

The Teaching Principle from *Principles and Standards for School Mathematics* elaborates the complexity of knowledge that an effective teacher needs: mathematical content, pedagogy, assessment strategies, and an understanding of students as learners. Teachers need a sound knowledge of the concepts, skills, and reasoning processes of mathematics to construct and achieve short- and long-term curricular goals. Teachers must develop a repertoire of strategies and pedagogical knowledge to guide instructional decisions. That repertoire must be coupled with a sound knowledge of students and how to further students' learning. Knowing that students make sense of mathematics in differing ways, teachers use their knowledge to create lessons that address, build on, and extend previous knowledge. No single "right way" exists to teach all mathematical topics in all situations; effective teachers balance their knowledge of mathematics, knowledge of pedagogical strategies, and knowledge of students to help students become independent mathematical thinkers.

No single "right way" exists to teach all mathematical topics in all situations; effective teachers balance their knowledge of mathematics, knowledge of pedagogical strategies, and knowledge of students to help students become independent mathematical thinkers.

Standard 1: Knowledge of Mathematics and General Pedagogy

Teachers of mathematics should have a deep knowledge of—

- sound and significant mathematics,
- theories of student intellectual development across the spectrum of diverse learners,
- modes of instruction and assessment, and
- effective communication and motivational strategies.

Elaboration

A goal of mathematics instruction is to enhance all students' understandings of both the concepts and the procedures of mathematics. To involve students in work that helps them deepen and connect their knowledge, teachers need specialized, content-specific knowledge for teaching mathematics. Specifically, teachers must themselves be experienced and highly skilled in the processes of problem solving, proof and

To involve students in work that helps them deepen and connect their knowledge, teachers need specialized, content-specific knowledge for teaching mathematics.

reasoning, communications, connections, and representation. To help students understand the connections within and across content domains and between mathematics and other disciplines, teachers must have a wide and deep knowledge of mathematics centered on the school mathematics they teach (Ball and Cohen 1999). Such knowledge includes—

> mathematical facts, concepts, procedures, and the relationships among them; knowledge of the ways that mathematical ideas can be represented; and knowledge of mathematics as a discipline—in particular, how mathematical knowledge is produced, the nature of discourse in mathematics, and the norms and standards of evidence that guide argument and proof.
>
> (National Research Council 2001, p. 371)

Equally important is how that knowledge is held by the teacher. Is it a collection of disconnected facts and algorithms, or is it a coherent, interconnected set of concepts that underlie and explain those facts (Ma 1999)?

Furthermore, teachers' mathematical knowledge should be framed within an understanding of human intellectual development. In planning for instruction and assessment, teachers must consider what they know about their students as well as what they know more generally about students from psychological, cultural, sociological, and political perspectives. Specifically, teachers should be well informed about issues of equity so they can ensure that their lessons contribute to a positive learning experience for all. The Equity Principle reminds us that all students can learn and do mathematics, that each one is worthy of being challenged intellectually, and that reasonable and appropriate accommodations should be made as needed.

But setting high expectations for all students is not enough. To elicit, explore, and critique students' mathematical thinking requires careful planning. Several studies have identified important ways in which teachers should attend to students' thinking in their planning and teaching by (1) understanding the mathematical concepts that will be developing during a lesson (e.g., Borko and Livingston 1989; Fernandez and Yoshida 2004; Lampert 2001; Leinhardt 1993; Leinhardt and Steele 2005; Livingston and Borko, 1990; Schoenfeld 1998; Schoenfeld, Minstrell, and van Zee 2000; Stigler and Hiebert 1999); (2) anticipating the variety of strategies students may use in solving a problem as well as the misconceptions or difficulties students may have (e.g., Borko and Livingston 1989; Fernandez and Yoshida 2004; Lampert 2001; Leinhardt 1993; Livingston and Borko 1990; Schoenfeld 1998; Schoenfeld, Minstrell, and van Zee 2000; Stigler and Hiebert 1999); and (3) asking questions to elicit students' thinking and advance students' understanding (e.g., Fernandez and Yoshida 2004; Schoenfeld 1998; Stigler and Hiebert, 1999). Attending to student thinking may also take a variety forms as it comes to life in classrooms.

Teachers must bring to the classroom a broad repertoire of instructional and assessment strategies. In particular, teachers need to understand how to engage students in various types of learning activities that are appropriate for both the students and the subject matter. How do you set up a productive cooperative learning experience? How do you facilitate a discussion or debate? What are the components of an effective guided exploration? How can you tell when a student truly understands a concept and when a student is merely mimicking an observed pattern or procedure? When and how do you help students make a transition from their own language or representations to more standardized terms and notations?

Assessment and instruction are often interwoven when making classroom decisions. To plan effectively, teachers must know how to assess their students' prior knowledge, how to determine what their students should learn next, and how they can intellectually challenge their students. Well-designed lessons afford teachers opportunities to learn about their students' understandings and encourage students to refine their understandings to accommodate new ideas.

Assessment and instruction are often interwoven when making classroom decisions.

To develop a motivational learning environment, teachers must understand what motivates their students. In particular, teachers must be familiar with their students' interests and abilities so they can help students see how mathematics emerges within those contexts. Similarly, when students are motivated by explorations using technology, manipulatives, or extended projects, teachers should find meaningful ways to engage students in those types of activities. Because students are willing to do what they believe they can do, teachers need to design learning experiences with various entry points so that students from diverse backgrounds can become engaged in the mathematics. Likewise, teachers must know how to select problems that are challenging enough to be interesting, yet not overwhelming, for their students.

Teachers should be knowledgeable of various verbal and nonverbal modes of communication. In particular, knowledge of technological applications can enhance a teacher's communication capabilities, providing a vehicle for dynamic graphical, numerical, and visual representations of mathematics concepts and skills. Likewise, hands-on and virtual manipulatives are powerful tools for representing mathematical concepts, thereby adding to a toolbox of representations from which students can communicate their mathematical understandings.

Standard 1: Knowledge of Mathematics and General Pedagogy

Vignette

In the first vignette, the teacher uses an exploratory, calculator-based task to spark students' mathematical thinking. While implementing the task, the teacher encourages her students to take intellectual risks by generating their own questions. As a result, the students generate ideas that relate to higher-level mathematics concepts. The teacher must draw on her own knowledge of those concepts in deciding how to respond to students' conjectures.

1.1—*Drawing on Mathematical Knowledge during Exploration Activities*

After recently completing a unit on multiplication and division, a fourth-grade class has just begun to learn about factors and multiples. Their teacher is having students use calculators as a tool for exploring that topic.

Using the automatic constant feature of their calculators (that is, that pressing 5 + = = = ... yields 5, 10, 15, 20, ... on the display), the fourth graders have generated lists of the multiples of different numbers. They have also used the calculator to explore the factors of different numbers. To encourage the students to deepen their understanding of numbers, the teacher has urged them to look for patterns and to make conjectures. She asked them, "Do you see any patterns in the lists you are making? Can you make any guesses about any of those patterns?"

Two students have raised a question that has attracted the interest of the whole class:

Are there more multiples of 3 or more multiples of 8?

The teacher encourages them to pursue the question, for she sees that it can engage them in the concept of multiples as well as provide a fruitful context for making mathematical arguments. She realizes that the question holds rich mathematical potential and even brings up questions about orders of infinity. "What do the rest of you think?" she asks. "How could you investigate this question? Go ahead and work on this a bit on your own or with a partner, and then let's discuss what you come up with."

Photograph courtesy of DeAnn Huinker;

The children pursue the question excitedly. The calculators are useful once more as the students generate lists of the multiples of 3 and the multiples of 8. Groups are forming around particular arguments. One group of children argues that there are more multiples of 3 because in the interval between 0 and 20 there are more multiples of 3 than multiples of 8. Another group is convinced that the multiples of 3 are "just as many as the multiples of 8 because they go on forever." A few children think there should be more multiples of 8 because 8 is greater than 3.

Although the teacher intends to revisit the students' conjecture another time, she redirects the conversation. She asks students whether they think the size of the number relates to how many factors it has. Some students excitedly form a new conjecture about factors: The larger the number, the more factors it has.

The teacher is pleased with the ways in which opportunities for mathematical reasoning are growing out of the initial exploration. The question asked by the two students has promoted mathematical reasoning, eliciting at least three competing and, to fourth graders, compelling mathematical arguments. The teacher likes the way in which students are making connections between multiples and factors. She also notes that students already seem quite fluent using the terms *multiple* and *factor*.

Although the class period is nearing its end, the teacher invites one group to present to the rest of the class their conjecture that the larger the number, the more factors it has. She suggests that the students record the conjecture in their notebooks and discuss it in class tomorrow. Pausing for a moment before she sends them out to recess, she decides to provoke their thinking a bit more: "That's an interesting conjecture. Let's just think about it for a second. How many factors does, say, 3 have?"

"Two," call out several students.

"What are they?" she probes. "Yes, Deng?"

Deng quickly replies, "1 and 3."

"Let's try another one," continues the teacher. "What about 20?"

After a moment, several hands shoot up. She pauses to allow students to think, and asks, "Natasha?"

"Six: 1 and 20, 2 and 10, 4 and 5," answers Natasha with confidence.

The teacher suggests a couple more numbers, 9 and 15. She is conscious of trying to use only numbers that fit the conjecture. With satisfaction, she notes that most of the students are quickly able to produce all the factors for each of the numbers she gives them. Some used paper and pencil, some used calculators, and some used a combination of both. As she looks up at the clock, one child asks, "But what about 17? It doesn't seem to work."

"That's one of the things that you could examine for tomorrow. I want all of you to see if you can find out whether this conjecture always holds."

"I don't think it'll work for odd numbers," says one child.

"Check into it," smiles the teacher. "We'll discuss it tomorrow."

The teacher deliberately decides to leave the question unanswered. She wants to encourage students to persevere and to not expect her to provide all the answers.

The teacher deliberately decides to leave the question unanswered. She wants to encourage students to persevere and to not expect her to provide all the answers.

In the second vignette, the teacher uses his knowledge of assessment techniques to diagnose student errors. The teacher realizes that asking students to explain their work may give him better insight into the children's thinking. As a result of his investigation, the teacher is able to develop a strategy to help his students connect their understanding of subtraction concepts with their paper-and-pencil work.

1.2—*Using Student Interviews to Identify Misconceptions*

The second graders have just finished working on addition and subtraction with regrouping. Although the students have been using a variety of models and written algorithms for subtraction, their teacher, Mr. Lewis, notices that many of them seem to "forget" to regroup in subtraction when using the traditional algorithm. Instead, they sometimes do this:

$$\begin{array}{r} 50 \\ -\,38 \\ \hline 28 \end{array}$$

Mr. Lewis wonders whether the students are being careless or whether something else is going on. He decides to sit down with the children one by one for a few minutes and have them talk through a couple of the problems and how they solved them. He thinks he may be better able to follow their thinking if they explain their steps as they work through a problem.

The teacher decides to use manipulative materials to gather some additional information about the students' understanding. He chooses a couple of problems from the test and asks the children to justify their answers using bundles of craft sticks. He discovers that most of them are not connecting the manipulatives work they did in class with their work on the written problems. When they use the craft sticks, they find that their paper-pencil answers do not make sense, and they revise them to match what they did with the sticks.

Mr. Lewis had assumed that if the students "saw" the concepts by actually touching the objects, they would understand. He now thinks that maybe he did not do enough to help them build the links between the concrete model and the algorithm. He starts wondering what he could do to make that connection clearer. He hypothesizes that perhaps the students know how to regroup but may not understand why we sometimes regroup or when regrouping is necessary. As a result of talking individually with students, Mr. Lewis concludes that he should revise his instructional plan to help students discover *why*, *when*, and *how* they should perform subtraction with regrouping. He decides to make up a set of examples in which regrouping is necessary for some and not for others. He plans to have the children discuss whether they would regroup in each example and how they would decide.

Closing Thoughts
Standard 1: Knowledge of Mathematics and General Pedagogy

Teachers must have a deep knowledge of mathematics on which to base their instructional decisions. However, knowledge of mathematical content alone is not sufficient to prepare teachers for the many challenges they face in the classroom. Knowledge of students' development, modes of instruction, and effective motivational and communication techniques all serve as resources for effective teaching practice.

Standard 2: Knowledge of Student Mathematical Learning

Teachers of mathematics must know and recognize the importance of—

- what is known about the ways students learn mathematics;
- methods of supporting students as they struggle to make sense of mathematical concepts and procedures;
- ways to help students build on informal mathematical understandings;
- a variety of tools for use in mathematical investigation and the benefits and limitations of those tools; and
- ways to stimulate engagement and guide the exploration of the mathematical processes of problem solving, reasoning and proof, communication, connections, and representations.

Elaboration

Effective mathematics teachers create opportunities for students to develop their mathematical understandings, competence, and interests (Lampert 2001). They actively engage students in tasks that enable them to see mathematics as a coherent and connected endeavor rather than as a series of disconnected rules and procedures that they must memorize. To support that kind of teaching, teachers need mathematics-specific pedagogical knowledge: knowledge of the important ideas central to their grade level, about representations that can most effectively be used in teaching those ideas, about the common misconceptions in learning those ideas, and about designing lessons that actively involve students in building their mathematical understanding (Shulman 1987).

Teachers rely on their pedagogical knowledge to make decisions about what to teach, when to teach it, and how to teach it. Specifically, teachers know how to assess their students' existing knowledge by carefully listening to students' explanations, by observing their manipulations of learning tools, and by analyzing their written work. One of the most valuable ways to identify and extend students' existing knowledge is to make use of problems that are set in real-world contexts with which students are familiar. In such settings, teachers can assess students' existing understandings as well as help students see the value of the mathematics they are learning. Teachers then draw on their knowledge and experience to connect with students intellectually and to help students revisit and refine their understandings. Mathematics learning is not a one-size-fits-all enterprise. Expert teachers know how to tailor experiences to fit the needs of individual students.

Mathematics learning is not a one-size-fits-all enterprise. Expert teachers know how to tailor experiences to fit the needs of individual students.

Teachers must know how to construct and sequence questions that engage students in the activities and help them focus on relevant aspects of the mathematics.

Mathematical ideas cannot be learned in a vacuum. Teaching mathematical topics as disconnected entities, or as a sequence of "tricks of the day," may lead to high quiz scores at the end of the week but rarely will lead to long-term understanding (Steen 1999). Rather, in-depth explorations of the relationships among representations and ideas help develop a more reliable and sustainable capacity to use, transfer, and understand mathematical ideas and procedures. Guiding such explorations requires extensive preparation and knowledge. Teachers must know how to construct and sequence questions that engage students in the activities and help them focus on relevant aspects of the mathematics. Teachers must realize when students have misinterpreted a situation and be prepared with questions that will help students discover their errors and rethink their strategies.

As students explore, investigate, and consider mathematical ideas, knowing when to interrupt and when to let students wrestle with ideas are important considerations for teachers. Although well-formulated solutions and clear explanations may be good indicators of understanding, merely listening to someone else's well-articulated ideas does not necessarily assure that the listener will understand or be able to offer a clear explanation when asked. As a result, teachers must be able to recognize appropriate moments to intervene as well as times when the students will benefit more by resolving challenges on their own (Hiebert and Wearne 2003).

Lessons that enable students to make connections among concrete, graphic, symbolic, and verbal representations often incorporate manipulatives, technology, and open-ended activities.

Lessons that enable students to make connections among concrete, graphic, symbolic, and verbal representations often incorporate manipulatives, technology, and open-ended activities. Such lessons require students to communicate using the language of mathematics as they reason through and solve problems. Because the lessons provide the stimulus for students to think about particular concepts and procedures, their connections with other mathematical ideas, and their applications to real-world situations, such lessons can help students to develop skills in contexts in which they will be useful. Teachers must have at their disposal, and be well versed in using, a variety of tools for generating, comparing, and connecting multiple representations. Teachers must select appropriate tools for the specific content at hand, must know how to exploit the conceptual advantages inherent in each tool, and must know how to address or avoid a tool's potential weaknesses—all this within the context of that teacher's classroom of students.

Although students should engage in a variety of mathematical processes throughout school, those processes may look very different in early elementary grades than they do in high school. As a result, teachers must know how to establish standards of reasoning that are appropriate for their own students and know how those standards may differ from what counts as a complete justification at other grade levels. Likewise, to effectively orchestrate a class discussion during which students share a variety of solution strategies, teachers must be aware of the most common strategies used to solve problems as well as the distinctive elements and connections among them.

Standard 2: Knowledge of Student Mathematical Learning

Vignette

In the first vignette, the teacher uses her knowledge of how children learn mathematics to develop a lesson that emphasizes conceptual understanding of division by a fraction. The teacher draws from existing curricular resources and modifies the tasks to better suit her pedagogical purposes.

2.1—*Modifying Resources to Meet Students' Needs*

Ms. Pierce is a third-year teacher in a large middle school. She uses a mathematics textbook, published about ten years ago, that her department strongly suggests she closely follow. In the midst of a unit on fractions with her seventh graders, Ms. Pierce is examining her textbook's treatment of division with fractions. Her previous two years of teaching experience and her teacher preparation background have caused her to be reflective in analyzing the goals and intent of textbook lessons. She is trying to determine this lesson's strengths and weaknesses and whether and how she should use the textbook as written to help her students understand division with fractions.

She notices that the textbook's emphasis is on the mechanics of carrying out the procedure ("dividing by a number is the same as multiplying by its reciprocal"). The text mentions that students "can use reciprocals to help" them divide by fractions and shows a few examples of the procedure. However, Ms. Pierce wants her students to understand what it means to divide by a fraction, not *simply* learn the mechanics of the procedure.

Ms. Pierce wants her students to understand what it means to divide by a fraction, not *simply* learn the mechanics of the procedure.

The picture at the top of one of the textbook pages shows some beads of a necklace lined up next to a ruler. This graphic is an attempt to represent that twenty-four 3/4-inch beads and forty-eight 3/8-inch beads are in an eighteen-inch necklace. Ms. Pierce notes that the graphic does represent what is meant by dividing by 3/4 or by 3/8, specifically related to the question "How many three-fourths or three-eighths are there in eighteen?" Still, when she considers what would help her students understand the problem, she does not think that this particular representation is adequate. She also suspects that students may not take the representation seriously, for they too often tend to perceive mathematics as memorizing rules rather than a means for understanding why the rules work.

Ms. Pierce senses that the idea of "using the reciprocal" is introduced almost as a trick, lacking any rationale or connection with the pictures of necklaces. Furthermore, division with fractions seems to be presented as a new topic, not connected with anything that the students might already know, such as division of whole numbers. Ms. Pierce is concerned that the textbook pages are likely to reinforce that impression. She does not see anything in the task that would emphasize the value of understanding why the procedure works, nor that would promote mathematical discourse.

Photograph by Susan Callum;

Thinking about her students, Ms. Pierce judges that the two pages under discussion require computational skills that most of her students already possess (i.e., being able to produce the reciprocal of a number and being able to multiply fractions) but that the exercises on the pages would not be interesting to them. Little on the two text pages would engage their thinking.

Looking at the pictures of the necklaces gives Ms. Pierce an idea. She decides she can use this idea, so she copies only the drawing. The necklace model is linear rather than circular, as are diagrams of a pie or pizza that are most often used to represent fractions. Ms. Pierce uses the linear model to help students develop varied representations. She also realizes that different representations make sense to different students.

To help students see connections between whole-number division and rational number division, Ms Pierce uses division models that can be used with whole numbers, rational numbers, or a combination of the two. She plans to include at least one picture that shows beads of whole-number length, for example, 2-inch beads (not shown here), before she shows pictures with beads of rational-number length, such as the ½-inch beads shown in the diagram. She will ask students to examine the pictures and try to write some kind of number sentence that represents what they see. For example, this 7-inch bracelet has fourteen 1/2-inch beads:

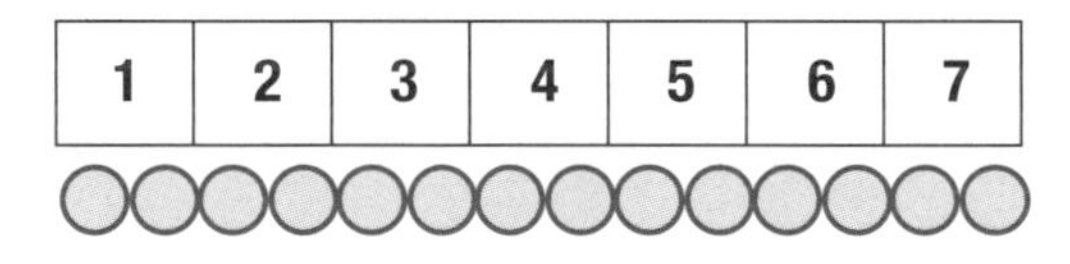

This situation could be represented as either $7 \div 1/2$ or 7×2. She will try to help students think about the reciprocal relationship between multiplication and division and the meaning of dividing something by a fraction or by a whole number. Then she thinks she could use some of the computational exercises on the second page of the textbook, but instead of just having the students compute the answers, she will ask them, in pairs, to write stories for each of about five exercises. Ms. Pierce knows that writing stories to go with the division sentences may help students focus on the meaning of the procedure.

She decides she will also provide a couple of other examples that involve whole-number divisors: $28 \div 8$ and $80 \div 16$, for example.

Ms. Pierce feels encouraged from her experience with planning this lesson and thinks that revising other textbook lessons will be feasible. Despite the fact that she is supposed to be following the text closely, Ms. Pierce now knows that she will be able to do so but still be able to adapt the text in ways that will significantly improve what she can do with her students this year.

Vignette

In the following vignette, a teacher assesses students' existing knowledge as the basis for a lesson in which students explore mathematical definitions and relationships among types of figures. Teacher-generated counterexamples serve as a powerful tool for helping students refine their language as they develop the components of a mathematical definition. When the students are ready, the teacher facilitates a discussion of how to further discriminate among figures that fit the definition.

2.2—*Helping Students Build on Informal Understandings*

Mrs. Logan is beginning an exploration of the properties of quadrilaterals for the purpose of creating a definition. She opens class by eliciting students' ideas, asking simply, "What do you know about quadrilaterals? How would you describe a quadrilateral?"

Several students respond in unison, "four sides," "four-sided figure."

Mrs. Logan reacts to exactly what the students say by making sketches that match students' descriptions. For example, Mrs. Logan draws

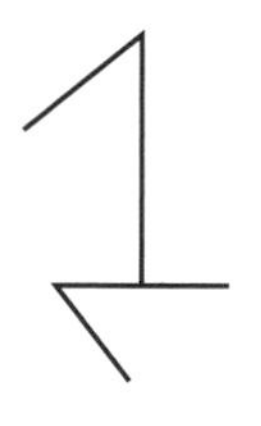

and asks, "Is this a quadrilateral?"

Students: No, it has to connect.

Mrs. Logan: Is this one?

Several students: No, it can't intersect like that.

Mrs. Logan continues drawing and asks, "So is this one?"

Student: It has to close.

Mrs. Logan: Okay, then is this one?

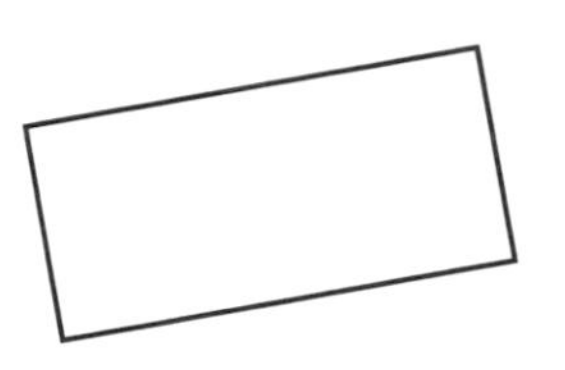

Students: Yes!

The teacher next asks a question designed to prompt students to identify specific characteristics of quadrilaterals on the basis of their discussion. Mrs. Logan pauses and then looks directly at the students: "I drew four examples. You said three of those didn't work. Can you explain what makes the difference?"

Several students volunteer elements that may become part of a definition of *quadrilateral*. Mrs. Logan lists their ideas—in their terms—on the board:

Quadrilaterals

4 points

4 segments

No more points intersect

Closed curve

Summarizing, Mrs. Logan tells them, "You really seem to have all the pieces. The definition of quadrilateral in our text is 'the union of four segments joined at their endpoints such that the segments intersect only at the endpoints.' Now, which of our figures fit with this definition?"

After a brief discussion during which they analyze their figures, Mrs. Logan continues, "And what are some special kinds of four-sided figures?"

Students call out a variety of names: *square, rectangle, kite, rhombus, parallelogram.*

Mrs. Logan: Another one you should have heard of before is *trapezoid.*

She draws several figures on the board, and asks, "Can you classify these and talk about them?"

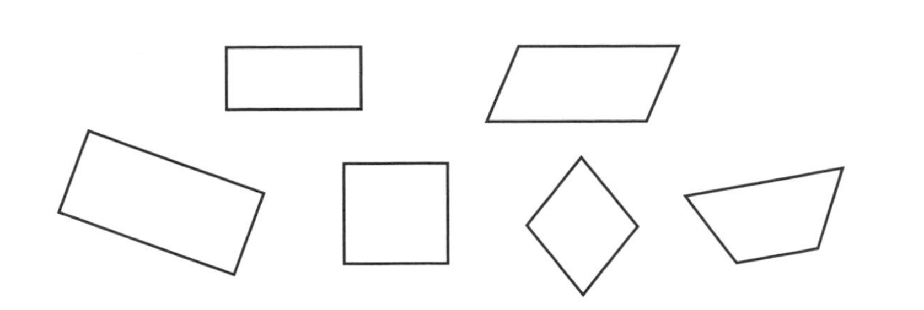

Students begin murmuring, in pairs and groups of three.

Mrs. Logan: Are there other special quadrilaterals that are not shown here?

One usually reticent student asks, "But isn't a square also a rectangle? I don't quite see how to classify these."

Instead of responding directly to his question, Mrs. Logan decides it is worth everyone's consideration and redirects the question to the group.

Mrs. Logan replies, "Why don't you put that to the rest of the class? See what they think."

The student repeats his question to classmates.

Mrs. Logan adds, "See if you can find a way to classify these shapes. Which shapes have which labels?"

The students resume their work. Some students begin making diagrams to represent the interrelationships among the types of quadrilaterals while others begin making tables. Mrs. Logan walks around, questioning students to get them to clarify what they are thinking.

Mrs. Logan asks one boy to explain his rather complicated chart. "And why is the rhombus there, with the parallelogram?"

Photograph by Mark Riley;

After class, Mrs. Logan contemplates the lesson. In general, she thinks it was a good start. Perhaps tomorrow—to help the students begin to construct some of the categories—she will engage them in a problem or activity in which they will have to sort quadrilaterals. She muses a bit about how to frame it in a way that will promote discourse and student understanding.

Closing Thoughts
Standard 2: Knowledge of Student Mathematical Learning

Beyond mathematics and general pedagogy, teachers need to draw on mathematics-specific pedagogy to best help their students develop both conceptual understanding and

The cycle of planning, implementing, and reflecting contributes both to students' learning and to teachers' continual improvement.

procedural proficiency. In particular, mathematics teachers must know how to actively engage their students in the mathematical processes of problem solving, reasoning, representing, communicating, and making connections. The cycle of planning, implementing, and reflecting contributes both to students' learning and to teachers' continual improvement.

Summary of the Knowledge Standards

The Teaching and Learning Principles of *Principles and Standards for School Mathematics* set goals for a new vision of the mathematics classroom. Departing from traditional practices, teachers use their knowledge of mathematics, pedagogy, and their students to develop lessons that foster students' learning and engagement. They assume that all students can learn and use significant mathematics, and they craft lessons to involve and challenge all students. How teachers implement their knowledge in the classroom is examined in the next section.

Creating an environment that supports and encourages mathematical reasoning and fosters all students' competence with, and inclination toward, mathematics should be one of a teacher's central concerns.

Implementation

Teachers must implement their knowledge about the teaching and learning of mathematics by choosing challenging tasks and facilitating meaningful mathematical discourse within a healthy and supportive learning environment. Creating an environment that supports and encourages mathematical reasoning and fosters all students' competence with, and inclination toward, mathematics should be one of a teacher's central concerns. The nature of such a learning environment is shaped by the kinds of mathematical tasks and discourse in which students engage. The tasks must encourage students to reason about mathematical ideas, to make connections, and to formulate, grapple with, and solve problems. In practice, students' actual opportunities for learning depend, to a considerable degree, on the kind of discourse that the teacher orchestrates. The nature of the activity and talk in the classroom shapes each student's opportunities to learn about particular topics as well as to develop her or his abilities to reason and communicate about those topics (Henningsen and Stein 1997).

Standard 3: Worthwhile Mathematical Tasks

The teacher of mathematics should design learning experiences and pose tasks that are based on sound and significant mathematics and that—

- engage students' intellect;
- develop mathematical understandings and skills;
- stimulate students to make connections and develop a coherent framework for mathematical ideas;

- call for problem formulation, problem solving, and mathematical reasoning;
- promote communication about mathematics;
- represent mathematics as an ongoing human activity; and
- display sensitivity to, and draw on, students' diverse background experiences and dispositions.

Elaboration

The tasks and activities that teachers select are mechanisms for drawing students into the important mathematics that composes the curriculum. Worthwhile mathematical tasks are those that do not separate mathematical thinking from mathematical concepts or skills, that capture students' curiosity, and that invite students to speculate and to pursue their hunches. Many such tasks can be approached in more than one interesting and legitimate way; some have more than one reasonable solution. Such tasks, consequently, promote student communication about mathematics by facilitating classroom discourse. They require students to reason about different strategies and outcomes, weigh the pros and cons of alternatives, and pursue particular paths. Worthwhile mathematical tasks can be drawn from print or electronic resources, or they can be created by teachers. Tasks can be designed to incorporate a wide range of tools available for teaching mathematics, such as Web-based resources, computer software, manipulative materials, calculators, puzzles, or interactive electronic devices.

Worthwhile mathematical tasks are those that do not separate mathematical thinking from mathematical concepts or skills, that capture students' curiosity, and that invite students to speculate and to pursue their hunches.

Teachers should maintain a curricular perspective, considering the potential of a task to help students progress in their cumulative understanding in a particular domain and to make connections among ideas they have studied in the past and those they will encounter in the future. For example, elementary school students are often taught that adding fractions and adding decimals are two entirely unrelated procedures. But a task in which students are asked to represent the sum of two decimal numbers as a sum of fractions that correspond to tenths, hundredths, thousandths, and so on, might help students recognize the connections between the two processes with respect to the combining of like terms.

Rather than ask students to memorize mathematical vocabulary, worthwhile mathematical tasks can be embedded in meaningful contexts that help students see the need for definitions and terms as they learn new concepts. For example, a method for helping students develop an understanding of function concepts is to introduce mathematical terms associated with functions and provide examples that illustrate those terms. However, a more worthwhile alternative might be to select a problem that involves a particular functional relationship, that underscores the need to consider the domain and range of the function, and that requires students to identify the meaning (or meaningless nature) of the function's inverse in the problem context.

Another content consideration is to assess what the task conveys about what is entailed in doing mathematics. Some tasks, although they might deal nicely with the concepts and procedures at hand, involve students in simply producing right answers. Other tasks can require students to speculate, to pursue alternatives, and to face decisions about whether their approaches are valid. For example, one task might require students to find means, medians, and modes for given sets of data. Another might require them to decide whether to calculate means, medians, or modes as the best measures of central tendency, given particular sets of data and particular claims they would like to make about the data, then to calculate those statistics, and finally to explain and defend their decisions. Like the first task, the second would offer students the opportunity to practice finding means, medians, and modes. Only the second, however, conveys the important point that summarizing data involves making decisions related to the data and the purposes for which the analysis is being used.

Tasks should foster students' sense that mathematics is a changing and evolving domain, one in which ideas grow and develop over time and to which many cultural groups have contributed.

Tasks should foster students' sense that mathematics is a changing and evolving domain, one in which ideas grow and develop over time and to which many cultural groups have contributed. Drawing on the history of mathematics can help teachers portray that idea. For example, teachers may ask students to explore alternative numeration systems or to investigate non-Euclidean geometries. Fractions evolved out of the Egyptians' attempts to divide quantities, such as four things shared among ten people. That fact could provide the explicit basis for a teacher's approach to introducing fractions.

Still another content consideration centers on the development of appropriate skill and automaticity. Teachers must assess the extent to which skills play a role in the context of particular mathematical topics. A goal is to create contexts that foster skill development even as students engage in problem solving and reasoning. For example, elementary school students should develop rapid facility with addition and multiplication combinations. Rolling pairs of dice as part of an investigation of probability can simultaneously provide students with practice with addition. Trying to figure out how many ways thirty-six desks can be arranged in equal-sized groups—and whether more or fewer groupings are possible with thirty-six, thirty-seven, thirty-eight, thirty-nine, or forty desks—prompts students to quickly produce each number's factors. As they work on the equal-groupings problem, students have concurrent opportunities to practice multiplication facts and to develop a sense of what factors are. Further, the problem may provoke interesting questions: How many factors does a number have? Do larger numbers necessarily have more factors? Does a number exist that has more factors than 36? Even as students pursue such questions, they practice and use multiplication facts, for skill plays a role in problem solving at all levels. Teachers of algebra and geometry must similarly consider which skills are essential, and why, and seek ways to develop essential skills in contexts in which they matter. What do students need to memorize? How can that memorization be facilitated?

Other factors to consider are students' interests, dispositions, and experiences. Teachers should aim for tasks that are likely to engage their students' interests. Not always, however, should such concern for "interest" limit the teacher to tasks that relate to the familiar, everyday worlds of the students. For example, theoretical or fanciful tasks that challenge students intellectually are also interesting. When teachers work with groups of students for whom the notion of "argument" is uncomfortable or at variance with community norms of interaction, teachers must consider carefully the ways in which they help students engage in mathematical discourse.

Knowledge about ways in which students learn mathematics is a basis for appraising tasks. The mode of activity, the kind of thinking required, and the way in which students are led to explore the particular content all contribute to the learning opportunity afforded by the task. Knowing that students need opportunities to model concepts concretely and pictorially, for example, might lead a teacher to select a task that involves such representations. An awareness of common student confusions or misconceptions about a certain mathematical topic would lead a teacher to select tasks that engage students in exploring essential ideas that often underlie those confusions or misconceptions. Understanding that writing about ideas helps clarify and develop understandings would lend attractiveness to a task that requires students to write explanations. Teachers' views about how students learn mathematics should be guided by research as well as by their own experience. Just as teachers can learn more about students' thinking from the tasks they pose to students, so, too, can they gain insights into how students learn mathematics. To capitalize on the opportunity, teachers should deliberately select tasks that provide them with windows through which to view students' thinking.

Knowledge about ways in which students learn mathematics is a basis for appraising tasks.

Although worthwhile mathematical tasks are rich with the potential for engaging students, their potential cannot be reached without thoughtful placement in the curriculum and careful implementation in the classroom. Therefore, teachers must take as much care in implementing tasks as selecting them. To maintain student engagement, teachers must select tasks that are demanding enough to require high-level thinking and then must allow students to wrestle with the central ideas. Although teachers must choose when to guide, support, and assist students, they must also take care not reduce the demands of the tasks so much that the important mathematics is diluted and the value of the tasks is diminished (Stein, Grover, and Henningsen 1996).

Although teachers must choose when to guide, support, and assist students, they must also take care not reduce the demands of the tasks so much that the important mathematics is diluted and the value of the tasks is diminished.

Standard 3: Worthwhile Mathematical Tasks

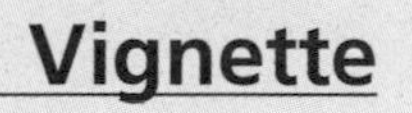

In the first vignette, the teacher selects a task to meet her curricular goals and the needs of her students. The teacher analyzes two tasks on the basis of the tasks' potential to intellectually stimulate her students and involve them in the processes of reasoning and problem solving.

3.1—*Selecting Intellectually Stimulating Tasks*

Mrs. Jackson is thinking about how to help her students learn about perimeter and area. She realizes that learning about perimeter and area entails developing concepts, procedures, and skills. Students need to understand that the perimeter is the distance around a region, that area is the amount of surface inside a region, and that length and area are measured in two different units. Students need to realize that although perimeter and area are not, in general, functions of one other, they are determined by common dimensional measurements: two figures with the same perimeter may have different areas, and two figures with the same area may have different perimeters. Students also need to be able to determine the perimeter and the area of a given region. At the same time, they should connect what they learn about the measurements of perimeter and area with what they already know about other measures, such as measures of volume or weight.

Mrs. Jackson examines two tasks designed to help upper-elementary-grade students learn about perimeter and area. She wants to compare what each has to offer.

Task 1

Find the area and perimeter of each rectangle:

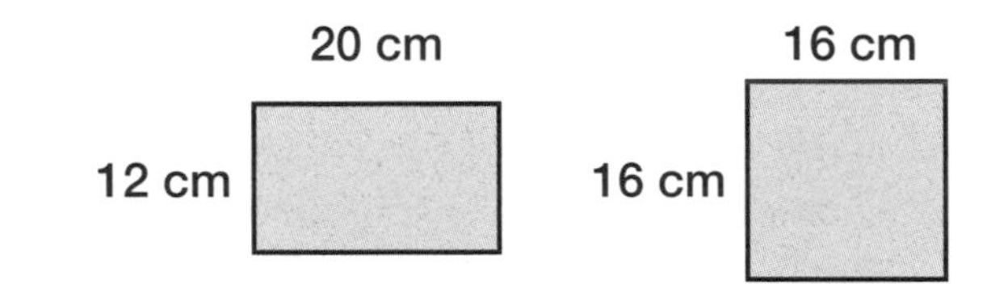

Mrs. Jackson decides that task 1 requires little more than correct application of formulas for perimeter and area. Nothing about the task requires students to make any conjectures about possible relationships between perimeter and area. This task is not likely to engage students intellectually; it does not entail reasoning or problem solving.

Task 2

> Suppose you had 64 meters of fence with which you were going to build a pen for your large dog, Bones. What are some different sized and shaped pens you can make if you use all the fencing? What does the pen with the least play space look like? What is the biggest pen you can make—the one that allows Bones the most play space? Which pen size would be best for running?

Mrs. Jackson believes that task 2 can engage students intellectually because it challenges them to search for something. Although accessible to even young students, the problem is not immediately solvable. Neither is it clear how best to approach it. A question that students confront as they explore the problem is how to determine that they have indeed found the largest or the smallest play area. To justify an answer and to show that a problem is solved are essential components of mathematical reasoning and problem solving.

Vignette

In the second vignette, the teacher uses one problem as a basis for a series of class activities related to translating an arrangement of cubes into a symbolic representation. Although the initial question the teacher poses can be solved by counting, he quickly makes use of the rich problem context to guide students to generate and give meaning to their own algebraic expressions.

3.2—*Making Sense of Algebraic Expressions*

Juan Rodriguez is beginning a unit on algebraic expressions with his ninth-grade students. He wants to use a problem that he saw discussed at a professional conference to determine whether it will help his students make more sense of algebraic expressions and better understand how different expressions can be equivalent.

Mr. Rodriguez distributes a bag of connectable unit cubes to each group of three students. He shows a large-image projection of the following stages for everyone to see.

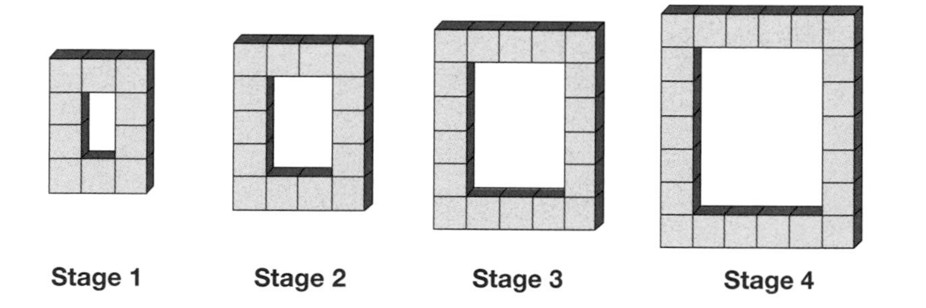

Mr. Rodriguez: In your groups, use your unit cubes to construct exact replicas of stage 1, stage 2, stage 3, and stage 4. Everyone at the table should check to make sure you have used the correct number of cubes in each stage.

The students enjoy the variety of activities that Mr. Rodriguez uses, so they quickly become engaged in the task. Mr. Rodriquez moves around the room to assure that students are focused on the task.

Mr. Rodriquez: On your mini white boards, please write the number of cubes that would be needed to construct the stage-5 figure. Raise your board so I can see your number.

The students discuss the question. Some students start building the stage-5 model while other students make calculations with paper and pencil. Eventually, all groups raise their boards.

Mr. Rodriguez: Sandra, can you tell us how your group got twenty-six?

Sandra: We used our cubes to build the next stage, and then we counted how many cubes it took to make it.

Mr. Rodriguez: Did any other groups build the stage-5 model to answer the question?

Students in several groups raise hands.

Mr. Rodriguez: Roger, how did your group decide how many cubes are required to build stage 5?

Roger: Well, each stage is like a rectangle with the middle cut out. And the next one is one cube longer and one cube wider than the stage before. So we knew that this one would have to be sort of a frame that was seven cubes by eight cubes.

Photograph by Terry McGowan;

Mr. Rodriguez: Very well explained, Roger.

The teacher chooses to not take the time to have students share alternative solution strategies because he knows that those will come up in response to other questions he has planned.

Mr. Rodriguez: Now, everyone, go back to you groups and discuss how many cubes it would take to build stage 11 in this sequence. Please raise your white boards when you have a number.

Mr. Rodriguez: I can see that every group has come up with fifty. Amanda, how did your group get fifty?

Amanda: Well, we didn't have enough cubes to build it, so we had to think of another way. We started by thinking of the four corners, then we added the rest of the cubes on the sides. So we added 4 to 11 and 11 and 12 and 12.

Mr. Rodriguez: Okay. Did another group do it differently? Raul?

Raul: We noticed that the length of the frame was 2 more than the stage number and the width was 3 more than the stage number. So we added two lengths and two widths, or two 13s and two 14s. That gave us 54, but we realized that we counted the corners twice, so we subtracted 4.

After asking several other students to explain their method of solving the problem, Mr. Rodriguez poses the following extension: "How many cubes would it take to build a frame, as you all have started to call these, at stage *n?* When you have an expression, please record it on the front board."

Students record several different expressions, including the following:

$$2n + 2(n + 1) + 4$$

$$2(n + 2) + 2(n + 3) - 4$$

$$n + n + (n + 3) + (n + 3)$$

$2n + 2(n + 3)$

$(n + 2) + (n + 2) + (n + 1) + (n + 1)$

$4n + 2 + 4$

$2(2n + 1) + 4$

$(n + 2)(n + 3) - (n)(n + 1)$

Mr. Rodriquez is pleased with the variety of expressions that are generated. He asks the class, "These all seem different. How can they all be correct? Maybe we should start by having you explain where these came from so we can see if they make sense. I'm especially intrigued by the last one. Bonnie, can you tell us how your group came up with that?"

Bonnie: We thought about the frame as a rectangle taken away from the middle of a bigger rectangle. So the outer rectangle has dimensions $(n + 3)$ by $(n + 2)$, and the inner rectangle has dimensions (n) by $(n + 1)$. Then we just subtracted the inner area from the outer area.

After discussing the rest of the student-generated expressions, Mr. Rodriguez plans to give the class another pattern of blocks and ask them a similar series of questions. He hopes that they will move to a generalization more quickly so that he can begin to ask questions that will help students focus on the concept of equivalence of algebraic expressions.

When reflecting on this class session, the teacher concludes that the exploration was worthwhile because all the students were quickly engaged in the task, it required that students give meaning to variables and expressions, and a valuable class discussion ensued.

Closing Thoughts
Standard 3: Worthwhile Mathematical Tasks

To engage students in challenging mathematics, teachers need to use worthwhile mathematical tasks. Those tasks must be rich in terms of content and processes. Tasks that promote communication and connections can help students see and articulate the value and beauty of mathematics. Teachers can enhance the value of existing materials by tailoring them to needs and interests of their students. By doing so, teachers can promote both motivation and equity in their classrooms.

Tasks that promote communication and connections can help students see and articulate the value and beauty of mathematics.

Standard 4: Learning Environment

The teacher of mathematics should create a learning environment that provides—

- the time necessary to explore sound mathematics and deal with significant ideas and problems;

- a physical space and appropriate materials that facilitate students' learning of mathematics;
- access and encouragement to use appropriate technology;
- a context that encourages the development of mathematical skill and proficiency;
- an atmosphere of respect and value for students' ideas and ways of thinking,
- an opportunity to work independently or collaboratively to make sense of mathematics;
- a climate for students to take intellectual risks in raising questions and formulating conjectures; and
- encouragement for the student to display a sense of mathematical competence by validating and supporting ideas with a mathematical argument.

Elaboration

The mathematics teacher is responsible for creating an intellectual environment in which serious engagement in mathematical thinking is the norm. More than just a physical setting with desks, bulletin boards, and posters, the classroom environment is suggestive of a hidden curriculum with messages about what counts in learning and doing mathematics. If teachers want students to learn to make conjectures, experiment with alternative approaches to solving problems, and construct and respond to others' mathematical arguments, then creating an environment that fosters those kinds of activities is essential (NCTM 2000).

A central focus of the classroom environment must be sense making.

A central focus of the classroom environment must be sense making (Hiebert et al. 1997). Mathematical concepts and procedures—indeed, mathematical skills—are central to making sense of mathematics and to reasoning mathematically. Teachers should consistently expect students to explain their ideas, to justify their solutions, and to persevere when they encounter difficulties. Teachers must also help students learn to expect and ask for justifications and explanations from one another. A teacher's own explanations must similarly focus on underlying meanings; something a teacher says is not necessarily true simply because he or she "said so."

Emphasizing reasoning and justification implies that students should be encouraged and expected to question one another's ideas and to explain and support their own ideas in the face of challenges from others (Lampert 2001). Teachers must assist students in learning how to do so. Teachers must create a classroom environment in which students' ideas, whether conventional or nonstandard, whether valid or invalid, are valued. Students need to learn how to justify their claims without becoming hostile or defensive. As teachers help students respect and show interest in the ideas of their classmates, students will be more likely to take risks in proposing their conjectures, strategies, and solutions.

Serious mathematical thinking takes time as well as intellectual courage and skills. A learning environment that supports problem solving must provide time for students to puzzle, to be stuck, to try alternative approaches, and to confer with one another and with the teacher. Furthermore, for many worthwhile mathematical tasks, especially those that require reasoning and problem solving, the speed, pace, and quantity of students' work are inappropriate criteria for "doing well." Too often, students have developed the belief that, if they cannot answer a mathematical question almost immediately, then they might as well give up. Teachers must encourage and expect students to persevere when they encounter mathematical challenges and invest the time required to figure things out. In discussions, the teacher must allow time for students to respond to questions and must also expect students to give one another time to think without interrupting or showing impatience (NCTM 2000).

Teachers must encourage and expect students to persevere when they encounter mathematical challenges and invest the time required to figure things out.

Students' learning of mathematics is enhanced in a learning environment that is a community of people collaborating to make sense of mathematical ideas (Hiebert et al. 1997). A significant role of the teacher is to develop and nurture students' abilities to learn with and from others—to clarify individual interpretations of definitions and terms with one another, to consider one another's ideas and solutions, and to argue together about the validity of alternative approaches and solutions. Various classroom structures can encourage and support such collaboration. Students may at times work independently, conferring with others as necessary; at other times students may work in pairs or in small groups. Students may use the Internet to research and collect data, use interactive geometry software to conduct investigations, or use graphing calculators to translate among different mathematical representations. Whole-class discussions are yet another profitable format. No single arrangement will work at all times; teachers should use various arrangements and tools flexibly to pursue their goals.

Students' learning of mathematics is enhanced in a learning environment that is a community of people collaborating to make sense of mathematical ideas.

Standard 4: Learning Environment

Vignette

The first vignette illustrates a learning environment in which students work independently and collaboratively, take responsibility for their own and one another's learning, and feel comfortable sharing ideas publicly. This environment fosters the development of student understanding about concepts and the connections among them.

4.1—*The Role of the Environment in Supporting Student Development*

Ms. Chavez is using a wireless learning system along with graphing calculators in her class and has her calculator connected to her LCD viewer. Her twenty-eight first-year algebra students, seated at round tables in groups of threes and fours, are working on a warm-up problem. The day before, they had had a test on functions. For the warm-up to today's class, Ms. Chavez has asked students to set up a table of values and use the table to create a graph of the function $y = |x|$.

She has chosen this problem as a way to introduce some ideas for a new unit on linear, absolute value, and quadratic functions. During the warm-up, students can be heard talking quietly to one another about the problem: "Does your graph look like a V-shape?" "Did you get two intersecting lines?" Walking around the room, Ms. Chavez listens to the students' conversations and uses her wireless system to capture the calculator screens and the work of students so she can get a feel for their attempts. After about five minutes, she signals that the time has come to begin the whole-group discussion.

A girl volunteers and carefully sketches her graph on a large dry-erase grid at the front of the room while the teacher displays the student's table of values using the LCD projector. As she does so, most students are watching closely, glancing down at their own graphs and tables, checking for correspondence. In this class, students are expected to communicate about mathematics. They also accept responsibility for helping others.

Another student suggests that they project Elena's graph, and the class watches as the graph appears on the projector screen. It matches the graph sketched by the first student, and the class cheers, "Way to go, Elena!"

Ms. Chavez then asks the class to sketch the graphs of $y = |x| + 1$, $y = |x| + 2$, and $y = |x| - 3$ on the same set of axes and write a paragraph that compares and contrasts the results with the graph of $y = |x|$. "Feel free to work alone or with the others in your group," she tells them.

After a few minutes, two students exclaim, "All the graphs have the same shape!"

A few other students look up. Another student observes, "They're like angles with different vertex points." "Then they're really congruent angles," adds his partner.

Ms. Chavez circulates through the classroom, listening to the discussions, asking questions, and offering suggestions. One group asks, "What would happen if we tried $|x - 3|$?" "Try it!" urges Ms. Chavez.

The students continue working, and the conversation is lowered to murmurs once again. Then the members of one group call out, "Hey, we've got something! All these graphs are just translations of $y = |x|$, just like we learned in the unit on geometry."

Ms. Chavez is pleased that her students seem to be making connections between this graphing activity and transformational geometry. "That's an interesting conjecture you have," remarks Ms. Chavez. She looks expectantly at the other students. "Do the rest of you agree?" They are quiet, many looking hard at their graphs. One student says, slowly, "I'm not sure I get it." Ms. Chavez has the students send their graphs, and all graphs are simultaneously displayed via the LCD projector. "Now, what do you think?" she asks.

A boy in the group that made the conjecture about translations explains, "Like, $y = |x| + 2$ is like $y = |x|$ moved up two spaces, and $y = |x| - 3$ is moved down three spaces. It's like what Louella said about them being like angles with different vertex points."

Ms. Chavez decides to prompt the class to pursue his idea. She asks whether anyone can graph $y = |x| + 4$ without first setting up a table of values and without using calculators. Hands shoot up. "Ooooh!" Scanning the class, Ms. Chavez notices that Lionel, who does not volunteer often, has his hand up. He looks pleased when she invites him to give it a try.

Lionel sketches his graph on the dry-erase board. Elena again enters the equation of the graph into the calculator, and the class watches as the graph is produced. The calculator-generated graph verifies Lionel's attempt. Again cheers erupt. Lionel gives a sweeping bow and sits down.

Ms. Chavez asks the students to write in their journals, focusing on what they think they understand and what they feel unsure about from today's lesson. Their journals give the teacher insights into students' thinking. They also offer students the opportunity to reflect on their understandings and feelings.

At the end of the period, Ms. Chavez distributes the homework that she has prepared. The worksheet includes additional practice on the concept of $y = |x| \pm c$ as well as something new, to provoke the next day's discussion: $y = |x \pm c|$.

Vignette

In the second vignette, students work together to solve problems. Sometimes they build on the solutions offered by classmates. The teacher has created an environment in which students expect to have to justify their solutions, not just give answers.

The teacher has created an environment in which students expect to have to justify their solutions, not just give answers.

4.2—*Encouraging Sense Making by Expecting Students to Reason*

A class of primary students has been working on problems that involve separating or dividing. The teacher, Laurie Morgan, is trying to give them some early experience with multiplicative situations at the same time that she provides them with contexts for deepening their knowledge of, and skill with, addition and subtraction. The students can add and subtract, but their understanding of multiplication and division is still quite informal. They have begun to develop some understanding of fractions, connected with their ideas about division. They have not yet learned any conventional procedures for dividing.

Today Ms. Morgan has given them the following problem:

If we make 49 sandwiches for our picnic, how many can each child have?

The teacher has selected this problem because it is likely to elicit alternative representations and solution strategies as well as different answers. It will also help the students develop their ideas about division, fractions, and the connections between them.

After they have worked for about ten minutes, first alone and then in small groups, Ms. Morgan asks whether the children are ready to discuss the problem in the whole group. Most, looking up when she asks, nod. She asks who would like to begin.

$$\begin{array}{r} 49 \\ -\,28 \\ \hline 22 \end{array}$$

Two girls go to the white board. They write the following problem:

One explains, "There are twenty-eight kids in our class, and so if we pass out one sandwich to each child, we will have twenty-two sandwiches left, and that's not enough for each of us, so there'll be leftovers."

The teacher and students are quiet for a moment, thinking about the proposed solution. Ms. Morgan looks over the group and asks whether anyone has a comment or a question about the solution. She expects the students, as members of a learning community, to decide whether an idea makes sense mathematically.

One boy says that he thinks their solution makes sense but that "9 minus 8 is 1, not 2, so it should be 21, not 22." He demonstrates by pointing at the number line above the white board. Starting at 9, he counts back eight using a pointer. The two girls ponder this demonstration for a moment. The class is quiet. Then one says, "We revise that. Nine minus 8 is 1." Ms. Morgan is listening closely but does not jump into the interchange. She wants students to respectfully question one another's ideas.

Another child remarks that he had the same solution as they did—one sandwich.

"Frankie?" asks Ms. Morgan, after pausing for a moment to look over the students. She remembers noticing his approach during the small-group time. Frankie announces, "I think we can give each child more than one sandwich. Look!" He proceeds to draw twenty-one rectangles on the white board. "These are the leftover sandwiches," he explains. "I can cut fourteen of them in half, and that will give us twenty-eight half-sandwiches, so everyone can get another half."

"I agree with Frankie," says another child. "Each child can have one and a half sandwiches."

"Do you have any leftovers?" asks the teacher.

"There are still seven sandwiches left over," says Frankie.

"What do the rest of you think about that?" inquires the teacher.

Several children give explanations in support of Frankie's solution. "I think that does make sense," says one girl, "but I had another solution. I think the answer is 1 plus 1/2 plus 1/4."

"I don't understand," Ms. Morgan says. "Could you show what you mean?"

Ms. Morgan encourages her students to take risks by bringing up different ideas. But raising questions is not enough. She also reminds students that they are expected to clarify and justify their ideas.

Closing Thoughts
Standard 4: Learning Environment

To capitalize on the value of worthwhile mathematical tasks, teachers and students must explore those tasks within a classroom environment that fosters intellectual growth and development. In such an environment, an objective of both teachers and students is to use mathematical tools and skills to make sense of problems, patterns, and contradictions. Participants use appropriate technology in such an environment to foster insight into mathematical situations. Students work independently or collaboratively to develop skills, make conjectures, and develop arguments within a mathematical community that values the contributions of all participants and defers to the authority of sound reasoning in the search for mathematical truth.

Standard 5: Discourse

The teacher of mathematics should orchestrate discourse by—

- posing questions and tasks that elicit, engage, and challenge each student's thinking;
- listening carefully to students' ideas and deciding what to pursue in depth from among the ideas that students generate during a discussion;
- asking students to clarify and justify their ideas orally and in writing and by accepting a variety of presentation modes;
- deciding when and how to attach mathematical notation and language to students' ideas;
- encouraging and accepting the use of multiple representations;
- making available tools for exploration and analysis;
- deciding when to provide information, when to clarify an issue, when to model, when to lead, and when to let students wrestle with a difficulty; and
- monitoring students' participation in discussions and deciding when and how to encourage each student to participate.

Elaboration

The discourse of a classroom—the ways of representing, thinking, talking, agreeing, and disagreeing—is central to what and how students learn about mathematics.

The Process Standards, most particularly the Communication Standard and the Proof and Reasoning Standard, make explicit reference to the role of discourse in the mathematics classroom (NCTM 2000). The discourse of a classroom—the ways of representing, thinking, talking, agreeing, and disagreeing—is central to what and how students learn about mathematics. Discourse represents both what the ideas entail and the ways ideas are exchanged: Who talks? About what? In what ways? What do people write? What do they record, and why? What questions are important? How do ideas change? Whose ideas and ways of thinking are valued? Who determines when to end a discussion? The discourse is shaped by the tasks in which students engage and the nature of the learning environment. The teacher's role is to initiate and orchestrate discourse and to use it skillfully to foster students' learning.

Like a piece of music, the classroom discourse has themes that are synthesized into a whole that has meaning. The teacher has a central role in orchestrating oral and written discourse in ways that contribute to students' understanding of mathematics. Students must be given opportunities to engage in making conjectures, to share their ideas and understandings, to propose approaches and solutions to problems, and to argue about the validity of particular claims; they must recognize that mathematical reasoning and evidence are the bases for discourse (NCTM 2000).

To effectively orchestrate mathematical discourse, teachers must do more listening, and students must do more reasoning.

To effectively orchestrate mathematical discourse, teachers must do more listening, and students must do more reasoning. Because many more ideas often come up than are fruitful to pursue at a given moment, teachers must filter and direct the students' explorations by picking up on some points and by leaving others behind. Doing so prevents student discourse from becoming too diffuse and unfocused. Such decisions depend on teachers' understandings of mathematics and of their students. In particular, teachers must make judgments about when students can figure things out individually or collectively and when they require additional input.

Knowledge of mathematics, knowledge of the curriculum, and knowledge of students should guide the teacher's decisions about the nature and the path of the discourse. Beyond asking clarifying or provocative questions, teachers should also, at times, provide information and lead students. Decisions about when to let students work to make sense of an idea or a problem without direct teacher input, when to ask leading questions, and when to tell students something directly are crucial to orchestrating productive mathematical discourse the classroom. Above all, the discourse should be focused on making sense of mathematical ideas and on using mathematical ideas sensibly in setting up and solving problems (Hiebert et al. 1997).

But how can teachers both stimulate and manage classroom discourse? Here are several suggestions:

1. Provoke students' reasoning about mathematics. Teachers do so through the tasks they provide and the questions they ask. For example, teachers should

regularly follow students' statements with "Why?" or "How do you know?" Doing so consistently, irrespective of the correctness of students' statements, is an important part of establishing a discourse centered on mathematical reasoning. In particular, when students are allowed to examine and critique incorrect solutions or strategies, counterexamples and logical inconsistencies can naturally surface. This process of analyzing solutions instead of relying on teachers to validate them can enhance students' abilities to think critically from a mathematical perspective.

Tasks and questions that stimulate discussion not only encourage active participation by students but also provide the teacher with ongoing assessment information. Cultivating a tone of interest when asking a student to explain or elaborate on an idea helps establish norms of civility and respect rather than criticism and doubt. The teacher must create an environment in which everyone's thinking is respected and in which reasoning and arguing about mathematical meanings is the norm.

2. Encourage the use of a variety of representations as well as explorations of how various representations are alike and different. Require students to write explanations for their solutions and to provide justifications for their ideas. Students should learn to verify, revise, and discard claims on the basis of mathematical evidence. Although students should learn to use and appreciate the conventional symbols that facilitate mathematical communication, teachers must value and encourage the use of a variety of tools and representations as students are learning to communicate their ideas. Various means for communicating about mathematics should be accepted, including drawings, diagrams, invented symbols, and analogies. The teacher should introduce conventional notation when students are ready to formalize their ideas and in a manner that helps students understand that using agreed-on terms and notations helps promote clear communication.

   As described in the Technology Principle, teachers should also help students learn to use calculators, computers, Internet resources, and other technological devices as tools for mathematical discourse. Given the range of mathematical tools available, teachers should often allow and encourage students to select the means they find most useful for working on or discussing a particular mathematical problem.

3. Monitor and organize students' participation. Teachers must be committed to engaging every student in contributing to the overall thinking of the class. Teachers must judge when students should work and talk in small groups and when the whole group is the most useful context. They must make sensitive decisions about how opportunities to speak are shared in the large group—for example, whom to call on, when, and whether to call on particular students who do not volunteer. Substantively, if discourse is to focus on making sense of mathematics, teachers must refrain from calling only on students who seem to have right answers or valid ideas to allow a broader spectrum of thinking to be explored in the discourse. By modeling respect for students' thinking and conveying the assumption that students make sense, teachers can encourage

students to participate within a norm that expects group members to justify their ideas.

Engaging every student in the discourse of the class requires considerable skill as well as an appreciation of, and respect for, students' diversity.

To facilitate learning by all students, teachers must also be perceptive and skillful in analyzing the culture of the classroom, looking out for patterns of inequality, dominance, and low expectations that are primary causes of non-participation by many students. Engaging every student in the discourse of the class requires considerable skill as well as an appreciation of, and respect for, students' diversity.

4. Encourage students to talk with one another as well as in response to the teacher. Whether working in small or large groups, students should be the audience for one another's comments—that is, they should speak to one another, aiming to convince or to question their peers. When the teacher talks most, the flow of ideas and knowledge is primarily from teacher to student. When students make public conjectures and reason with others about mathematics, ideas and knowledge are developed collaboratively, revealing mathematics as constructed by human beings within an intellectual community. Students, who are accustomed to the teacher's doing most of the talking while they remain passive, need guidance and encouragement to participate actively in the discourse of a collaborative community. Some students, particularly those who have been successful in more traditional mathematics classrooms, may be resistant to talking, writing, and reasoning together about mathematics. Time and patience are required to cultivate a student-centered environment for discourse. Eventually, the payoff in student participation and learning is worth the investment.

Eventually, the payoff in student participation and learning is worth the investment.

Standard 5: Discourse

Vignette

In the first vignette, the teacher facilitates a discussion about counting with young children. The choice of task creates an opportunity for reasoning and problem solving. The teacher's consistent urging of students to explain and to justify their answers invites many students into the conversation and encourages the sharing of a variety of solution methods.

5.1—*Only the Nose Knows, but the Children Can Reason!*

Ms. Nakamura has done a lot of number work with her kindergarten class this year, and she is pleased with the results. Now, near the end of the year, the class has been investigating patterns in the number of various body parts in the classroom—how many noses or eyes, for example, are present among the children in the class.

Earlier that week, each child had made a nose out of clay. Ms. Nakamura opens the discussion by revisiting that project. She asks, "And how many noses did we make?"

Becky (points to her nostrils): Two of these.

Teacher: But how many actual noses?

Anne: Twenty-nine.

Teacher: Why? Why were there twenty-nine noses?

Adam: Because every kid in the class made one clay nose, and that is the same number as kids in the class.

Teacher (pointing to her nostrils): Now Becky just said—remember what these are called?

Children: Nostrils!

Teacher: So were there twenty-nine nostrils?

Pat: No, there were more.

Gwen: Fifty-eight! We had fifty-eight nostrils!

Teacher: Why fifty-eight?

Gwen: I counted.

Felice: If we had thirty kids, we would be sixty. So it is fifty-nine 'cause it should be one less.

Teacher: Can you explain that again?

The teacher probes Felice's answer even though her approach goes beyond what many of the children are trying to do at this point.

Felice: It's fifty-nine because we don't have thirty kids, we have twenty-nine, so it is one less than sixty.

Teacher: What does anyone else think?

Adam: I think it is fifty-eight. Each kid has two nostrils. So if sixty would be for thirty kids, then it has to be two less: fifty-eight.

The teacher solicits other students' reactions instead of showing them the right answer. Her tone of voice and her questions show the students that she values their thinking.

Lawrence: But Felice says thirty kids makes sixty …

Felice: No! Adam makes sense. Fifty-eight.

The teacher moves on, asking, "What else do you think we have on our bodies that would be more than twenty-nine?"

The teacher's question challenges students to think it is open-ended; more than one right answer exists.

Graham: More than twenty-nine fingers.

Teacher: More than twenty-nine fingers? Why do you think so?

Graham: Because each kid, we have ten fingers.

Ricky: More than twenty-nine shoes.

Teacher: More than twenty-nine shoes. And what are those shoes covering?

Ricky: Your feet.

Sarah: Ears.

Beth: More than twenty-nine legs.

Ms. Nakamura tells the children, "You did some good thinking today!" The teacher chooses to comment on the children's thinking instead of their behavior.

Ms. Nakamura tells the children that they now are to work on a picture: "Choose some body part, and draw a picture of how many of those we have in our class and how you know that." She directs the children back to their tables, where she has laid out paper and cans of crayons.

Vignette

In the following vignette, the teacher encourages the use of multiple representations to help students construct understandings. The students use objects and symbols to help convey their meaning. The teacher eventually introduces standard notation and helps students make connections between their own and other representations.

5.2—*Making the Transition from Student-Invented to Standard Representations*

Mr. Johnson has presented his first-grade class with several pairs of numbers and asked them to decide which number in each pair is greater and to justify their responses. He has also been encouraging them to find ways to write their comparisons. The students know that the teacher expects not just answers but also reasons.

Ben: I think 5 is greater than 3 because [walks to the board and sticks five magnets up and then carefully sticks three magnets in another row]).

Mr. Johnson asks whether that explanation makes sense to other people. The children nod. He asks whether anyone wants to show how you would write Ben's idea down. The teacher encourages students to use symbols to represent and communicate about ideas.

Kevin, up at the board, writes this expression:

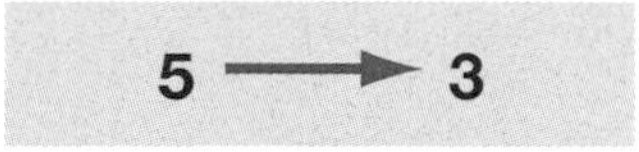

Next, Betsy writes the following:

Mr. Johnson: Can you explain what you were thinking? Kevin?

Kevin explains that his arrow shows that 5 is more than 3 because the bigger number "can point at" the smaller one.

Mr. Johnson asks Betsy to explain hers, and she says that she thinks you should just circle the smaller one.

Photograph courtesy of Deborah A. Niezgoda and Patricia S. Moyer-Packenham; all rights reserved

The teacher accepts more than one way of representing the idea of comparing numbers by using symbols; both are nonstandard but sensible. However, Mr. Johnson challenges his students by posing a question that requires students to invent a means of recording an idea.

Mr. Johnson: What if the two numbers you were comparing were 6 and 6? What would you do? How would you use your symbols for comparing numbers to write that?

Several seconds pass. The teacher gives students time to think before responding. He doesn't repeat the question or call on children; he is patient and silent. Eventually, Ruth shoots her hand in the air. Several others also have their hands up.

Ruth: You could draw an arrow to both of them.

Annie: You could circle both of them because they are the same.

Jimmy: You shouldn't mark either one, either way. They are not greater or less. They are the same.

Mr. Johnson nods at their suggestions. He writes an equals sign (=) on the board and explains that this is a symbol that people have invented for the ideas the children have been talking about.

The teacher connects the students' approaches and reasoning with the conventional notation. Because the students have thought about what it means for two numbers to be equal, they are ready to learn how that relationship is conventionally represented. In this situation, the notation follows the development of the concept in a meaningful context.

Rashida: That's like what Ruth said.

Ruth beams, and Annie calls out, "It's like mine, too."

Vignette

In the third vignette, a teacher introduces a collaborative project. The teacher monitors small-group discussion and planning. He offers advice and resources when appropriate.

5.3—*Letting the Discourse Happen: Monitoring Collaborative Groups*

Mr. Cohen's class of high school students is working in small groups on projects that involve collecting, organizing, and interpreting data. The teacher has posed a task that gives the students an opportunity to develop their understanding of *sample* and *population*. The task is also intended to help students extend their ability to use statistics to reason about real-world situations.

The class first discussed possibilities for projects. They identified questions that they wanted to pursue, such as finding the average number of hours per week that high school students work.

In one group, a student had recently read a newspaper article on changes in the popularity of first names since 1925. The group has decided to investigate the most common boys' and girls' first names among students their age in their community. Are Michael, James, and Robert still the most common boys' names, as they were in 1925, 1950, and 1975, respectively? They are curious about what has happened with girls' names, because, according to the article, girls' popular names seem to change more often. They also wonder how their community's ethnic diversity affects the pattern of names.

Mr. Cohen moves around the room to the different groups. He stops to listen, offer suggestions, and verify that the group members are listening to one another and working together. The teacher is providing time for students to grapple with the data-collection aspects of their projects. The group that is working on the names study has decided to sample the high school population in the city and is discussing the best way to go about doing so. They intend to compare their results with national data available from the Social Security Administration Web site, a data source that was provided in the newspaper article.

Photograph courtesy of Joseph Dakoske; all rights reserved

John: Let's choose three of the high schools and then write to them and ask for a list of the students enrolled in the school. We can sample names at random from those lists.

Jenny: But how will we pick the three schools? And why is three a good number to pick?

John: It seemed like enough out of all the schools in our city if we were careful to include one of the schools that has more kids from different backgrounds, because we want to make sure our sample has lots of different kinds of names, just like there are around here.

Anna: I think we should try to figure out about how many high school kids there are in the whole city and then pick a size for our sample based on that.

John (nodding): I guess that makes sense. How are we going to figure that out, though?

Maria: And then how big would our sample have to be to be big enough? We want to be pretty sure that our sample tells us something about all the kids in high school here.

Mr. Cohen is standing by the group. These students seem to have developed the disposition to question one another, and they seem accustomed to thinking through problems together.

These students seem to have developed the disposition to question one another, and they seem accustomed to thinking through problems together.

Mr. Cohen tells the students that their discussion so far is productive, that they are dealing with some important questions for their project. He suggests a source that might help them think about the question of how many students they need to have in their sample. He also tells them that the school administration office would have a list of all the high schools and how many students attend each of them. He asks if they would like him to call and ask for that list. They say that they would. He asks what they are going to do once they get the list.

Closing Thoughts
Standard 5: Discourse

Although some level of communication occurs in every mathematics classroom, the type and the amount of participation in that communication are varied. In one-way communication from teacher to students, students have very little opportunity to

Careful monitoring of student discourse and posing questions that challenge students to refine and reorganize their ideas improve the quality of the discourse as well as students' developing understandings.

grapple with mathematical ideas within an intellectual community. When discourse patterns include student-to-student discourse and student-to-teacher discourse, students find opportunities to translate their thoughts into verbal, symbolic, and graphical representations. By using those representations, students develop facility with them and attach meaning to them. Careful monitoring of student discourse and posing questions that challenge students to refine and reorganize their ideas improve the quality of the discourse as well as students' developing understandings.

Summary of the Implementation Standards

Implementation of teachers' knowledge is evident in the ways teachers choose tasks, the means by which they create a supportive and challenging environment for learning, and the tools they use to orchestrate discourse in the classroom. Because the teacher is responsible for shaping and directing students' activities so that they have opportunities to engage meaningfully in mathematics, the tasks in which students engage must encourage them to reason about mathematical problems. By expecting students to participate, listen respectfully to one another, present their ideas, and pose questions to the teacher and to peers, the teacher establishes an environment that nurtures the learning of mathematical processes and concepts and the development of skills.

The discourse of the mathematics class reflects messages about what it means to know mathematics, what makes something true or reasonable, and what doing mathematics entails. It is central to both what students learn about mathematics and how they learn it. Therefore, the discourse of the mathematics class should be founded on mathematical ways of knowing and ways of communicating. Although a teacher may be more of a "guide on the side" than a "sage on the stage," the teacher is the central element in fostering worthwhile mathematical discourse within the classroom community.

Teachers' skills in developing and integrating the tasks, discourse, and environment in ways that promote students' learning are enhanced through the thoughtful analysis of their instruction. That process of analysis is the focus of the next, and final, group of standards in this chapter.

Analysis

A central question to which teachers must be prepared to respond is "How well are the tasks, discourse, and learning environment working to foster the development of students' mathematical proficiency and understanding?"

Trying to understand as much as possible about the various elements that affect the learning of each student is essential to good teaching.

Trying to understand as much as possible about the various elements that affect the learning of each student is essential to good teaching. Teachers must monitor classroom life using a variety of strategies and focusing on a broad array of dimensions of mathematical competence, as described in the Assessment Principle of *Principles and Standards for Schools Mathematics* (NCTM 2000). The information gained from formative assessment should guide instruction. What do students seem to understand well, and what do they appear to understand only partially? What connections do they seem to be making? What dispositions toward mathematics do they convey? How does the class work together as a learning community making sense of mathematics? What teachers learn from their analysis should be a primary source of information for planning and improving instruction.

Standard 6: Reflection on Student Learning

The teacher of mathematics should engage in ongoing analysis of students' learning by—

- observing, listening to, and gathering information about students to assess what they are learning

so as to—

- ensure that every student is learning sound and significant mathematics and is developing a positive disposition toward mathematics;
- challenge and extend students' ideas;
- adapt or change activities while teaching;
- describe and comment on each student's learning to parents and administrators; and
- provide regular feedback to the students themselves.

Elaboration

Assessing students and analyzing instruction are fundamentally interconnected. Mathematics teachers should monitor students' learning on an ongoing basis to assess and adjust their teaching. Observing and listening to students during class can help teachers, on the spot, tailor their questions or tasks to provoke and extend students' thinking and understanding. Students' dispositions toward mathematics—their confidence, interest, enjoyment, and perseverance—constitute another important dimension for teachers to monitor. Teachers have the responsibility of describing and commenting on students' learning to administrators, to parents, and to the students themselves so as to guide each student to a better understanding of his or her personal learning style.

Teachers must assess the skills, knowledge, and conceptual and procedural understanding of their students. They must also assess the development of students' ability to reason mathematically—to make conjectures, to justify and revise claims on the basis of mathematical evidence, and to analyze and solve problems. Paper-and-pencil tests, although one example of a useful means for judging certain aspects of students'

The value of good questions, such as those that ask "Why?" "How do you know?" and "What if?" should be underscored here as a way to stimulate the kind of activities and discussion that can open a window into students' thinking.

mathematical knowledge, cannot give teachers the insights they need about students' understandings to make instruction as effectively responsive as possible. Teachers need information gathered in a variety of ways and using a range of sources. Observing students participating in a small-group discussion may contribute valuable insights about their abilities to communicate mathematically. Interviews with individual students can complement that information and also yield information about students' conceptual and procedural understanding. Students' journals are yet another source that can help teachers appraise their students' development. Teachers can also learn a great deal by closely watching and listening to students during whole-group discussions. The value of good questions, such as those that ask "Why?" "How do you know?" and "What if?" should be underscored here as a way to stimulate the kind of activities and discussion that can open a window into students' thinking. Finally, teachers need to analyze information garnered from both formative and summative assessments and use the resulting insights into students' understanding to guide their instructional decisions.

Standard 6: Reflection on Students' Learning

Vignette

In the first vignette, a teacher uses a nonroutine problem to assess students' understanding of quadratic equations as well as the development of their dispositions toward mathematics. The teacher gathers information about their understanding from listening to students as they solve the problem and by reading their journal entries related to the activity. As a result of the assessment, the teacher decides to revise his plans for subsequent lessons.

6.1—*Using a Challenging Problem to Assess Student Understanding*

Toward the end of a unit on quadratic equations, Mr. Santos has decided to assess his algebra students' use of problem-solving processes and their ability to make mathematical connections, both among ideas in the unit and between those ideas and concepts learned earlier. To do so, he chooses a problem from the 1988 NCTM Yearbook, *The Ideas of Algebra, K–12* (Coxford 1988, p. 19) He notices that it is headed "Can Your Algebra Class Solve This?" and realizes that it is likely to be a tough one for them, but he expects that as a nonroutine problem, it will serve as an alternative means of assessment. He hopes to gain insight into the students' learning and development of mathematical disposition. The problem is stated as follows:

Find all real values of x that satisfy

$$(x^2 - 5x + 5)^{x^2 - 9x + 20} = 1$$

Mr. Santos decides to ask students to work on the problem in pairs while he circulates among as many of the pairs as he can, monitoring their progress. He uses a checklist with students' names on it as an easy means of recording observations about students' thinking, approaches, and patterns of working.

The first pair of students he visits, Alan and Bettina, groan, "This is really going to be gross!" "Look, it's got two different quadratics in the same equation!" "Yeah, it's not fair. He never gave us such a complicated one before!" "Oh, well," Bettina says, "we might as well get started. Let's factor $x^2 - 9x + 20$ and see what we get." When they find that $(x - 4)$ and $(x - 5)$ are the factors, Alan says, "Well, I guess that's it, $x = 4$ and $x = 5$ must be the answers."

Bettina does not seem certain about Alan's assertion. "What about this other quadratic? Don't we have to check that it works there, too?" she asks. "Oh, yeah," agrees Alan, "you check 5, and I'll check 4." So they substitute 4 and 5 into $x^2 - 5x + 5$ and find that they get 1 for $x = 4$ and they get 5 for $x = 5$. Alan says, "I get 1 like I'm supposed to," but Bettina says, "I don't get 1, I get 5." Bettina's result puzzles them.

As they look at the problem together, Alan says, "We need to use both quadratics together," and Bettina chimes in, "Yeah, it's this to that power." Evaluating the entire expression, they find that for $x = 4$ they get 1^0 and for $x = 5$ they get 5^0. They comment that "It's 1 either way; anything to the 0 power is 1."

Alan leans back, seeming confident that they have solved the problem. Bettina, still engaged with the problem, says, "Hey, look, if this is 1, then the exponent could be anything. Can we use that?" The teacher notices with pleasure that Bettina seems persistent, reflective, and on the lookout for additional solutions.

Taking up Bettina's question, Alan points at the base, $x^2 - 5x + 5$, and says, "Okay, you mean we should see if any other values of x could make this part equal to 1?"

Out of the corner of his eye, Mr. Santos notices a pair of students clowning around by the window. He hears them laughing and sees them pushing one another playfully. He approaches them and asks, "What's up?"

"No way we can do this problem, Mr. S," says Diarra.

"And, besides, who *cares?*" adds Tommy.

Mr. Santos guesses that part of their frustration is that the problem looks too complex. He invites them to try the problem by putting in some numbers.

"How about 1?" suggests Diarra, giggling.

"Yeah," agrees Tommy.

When they try 1, they are surprised to see that it works out.

Photograph by Dave Ebert;

"Hey, this is easy, man!" exclaims Tommy. At this remark, other students crowd around.

"Are there other solutions?" asks Mr. Santos, relieved that the students are becoming engaged.

"I'll try 2," volunteers one. Others are trying other numbers. As he walks away, Mr. Santos hears another burst of excitement as a pair of students discovers that 2 also works. He also hears a groan from a student who has tried 0.

Looking around the classroom, Mr. Santos notices a pair of students, Geri and Linnea, using a graphing calculator. When he goes over to them, they tell him that they have graphed the functions $y = 1$ and

$$y = (x^2 - 5x + 5)^{(x^2 - 9x + 20)}$$

and are now zooming in to identify the points of intersection. When he asks them to explain what they have been doing, they say that they decided from the beginning to use a graphing calculator. They describe moving from graphing the two quadratics separately to using the exponentiation key and graphing the whole function at once. Linnea says, "We finally realized that what we had here was a polynomial to a polynomial power."

Mr. Santos asks them about the section of discontinuity on the graph and whether their "picture" represents a complete graph. He suddenly realizes that this pair of students has given him additional insight into this problem, and he makes a mental note to change his lesson plans for later in the week. He will bring in the demonstration computer so that the whole class can participate in further discussion on using technology to solve this problem.

Mr. Santos looks around the room for another group to visit and notices another pair, Peter and Ona, lounging with nothing to do. "How are you two doing?" he inquires pleasantly.

"Great!" Peter replies, "We got the answer; it's 4 and 5." They show Mr. Santos how they did it. They have used an approach similar to the one used by Alan and Bettina.

Mr. Santos asks, "Didn't you just say that when $x = 4$, you got this polynomial [pointing to the base, $(x^2 - 5x + 5)$] to be equal to 1?" He pauses, hoping that they will notice the importance of the base having the value 1.

After some consideration, Peter says, "Yes, but we were worried more about the exponent being 0; but if the base is 1, the exponent wouldn't have to be 0." Ona says,

"Okay, let's see if we can solve $x^2 - 5x + 5 = 1$. So they set out to factor $x^2 - 5x + 5$, ignoring the fact that it is not set equal to 0.

Mr. Santos glances at his watch and sees that the period is almost over. He decides to end the class by reminding the students to write their journal entries for the day. They are to record the problems and successes they encountered during the period, any new insights, and anything that stood out to them about other students' arguments or solutions in class.

Vignette

In the following vignette, a teacher discusses students' progress with parents. The teacher draws on a system that she has developed for collecting and analyzing information about her own teaching to share information about students' understanding with parents.

6.2—*Documenting Student Work to Monitor and Report Progress*

Ms. Lundgren has been trying to change her approach to teaching mathematics so that students are not only developing procedural fluency but are also learning to reason and communicate about mathematics, to make sense of mathematical ideas, and to make connections. She believes that she has been successful in moving the discourse of her classroom away from solely a focus on right answers and the teacher as authority. The teacher knows that she must find some ways of documenting and assessing what students are learning, especially in view of her new goals for them.

Although difficult to accomplish, she has also been devising better mathematical tasks, she thinks. With the help of the other fifth-grade teacher, Ms. Lundgren has also come up with some ways of keeping track of what students are learning. Today she is meeting with parents to discuss their children's report cards, and she has decided to draw on her new records for these conferences. The teacher wants the parents to understand both what their children are doing and what is being held as important in her mathematics class.

When she is meeting with Mrs. Byers, Stacy's mother, Ms. Lundgren wants to show her how Stacy is making connections in division. Looking at her card on Stacy, Ms. Lundgren tells the mother that Stacy was able to explain how, for 28 ÷ 8, 3 r 4 was the same answer as 3.5 (a quotient obtained on the calculator) but also how the two answers differed. Ms. Lundgren, having made a note of it, opened Stacy's mathematics journal to the page where Stacy had worked this problem out. Then, referring to the index card again, Ms. Lundgren shows Stacy's mother all the ways that Stacy found to represent 8 ÷ 1/2 in her journal. In addition to showing Stacy's procedural fluency and level of mathematical understanding, she also wants to talk with Mrs. Byers about Stacy's disposition toward mathematics, Ms. Lundgren refers to a chart she is keeping on her students' mathematical attitudes. With this chart, she has periodically made notes to herself. She has also had her colleague next door come in and observe once a month and make notes on the chart for her. Mrs. Byers finds all these specific examples very useful and comments that she thinks that what Ms. Lundgren is trying to do in mathematics is great and she wishes she had had a mathematics class like this when she was in school.

Closing Thoughts

Standard 6: Reflection on Student Learning

The essential factor in growth and improvement in teaching is the analysis of lesson outcomes both during and after each lesson.

Although well-planned lessons that incorporate worthwhile tasks within a supportive environment go a long way toward successful teaching and learning, students do not always respond as anticipated and activities often lead classes in unintended directions. The essential factor in growth and improvement in teaching is the analysis of lesson outcomes both during and after each lesson. When students pose questions that may lead the class away from the intended lesson objectives, teachers must make instantaneous decisions about the relative merits of following an unintended path. When students appear to be stuck without much hope of making further progress, teachers must decide whether they should intervene and what type of help to offer. In some instances, entire lesson plans must be scrapped in favor of more- or less-challenging activities. In addition to on-the-spot decision making, teachers must reflect on student outcomes by way of assessing students' work in its various forms. This continual monitoring requires careful listening and data collection to yield the most useful information for improving practice.

Standard 7: Reflection on Teaching Practice

The teacher of mathematics should engage in ongoing analysis of teaching by—

- reflecting regularly on what and how they teach;
- examining effects of the task, discourse, and learning environment on students' mathematical knowledge, skills, and dispositions;
- seeking to improve their teaching and practice by participating in learning communities beyond their classroom;
- analyzing and using assessment data to make reasoned decisions about necessary changes in curriculum; and
- collaborating with colleagues to develop plans to improve instructional programs.

Elaboration

As they monitor students' understandings of, and dispositions toward, mathematics, teachers should reflect on the nature of the learning environment they have created, on the tasks they have been using, and on the kind of discourse they have been fostering. They should seek to understand the links between those factors and what is happening with their students. Have the selected tasks challenged students and motivated them to stick with difficult problems? Has the scaffolding of certain ideas helped build prerequisite skills? Has the classroom environment helped students become more confident, more comfortable with sharing ideas, or better able to ask relevant questions? Has the discourse included mathematical discussions among students as well as between the teacher and students? Have the activities and questions caused students to increase their ability to think mathematically, as well as to arrive at correct mathematical solutions? Are the students drawing on a variety of tools, technologies, representations, and strategies as they solve problems?

Teachers need to analyze continually what they are seeing and hearing and explore possible interpretations of that information. What do such insights suggest about how the environment, tasks, and discourse could be enhanced, revised, or adapted to help all students learn? How should this information affect short- and long-term planning?

To ensure that teachers are assessing what they have taught, assessments should reflect both the content and form of instruction. If reasoning and problem solving are the foci of instruction, they should also be the foci of assessment. Formative assessments, such as journal entries, quizzes, interviews, or even classroom observations, should be an ongoing part of teachers' practice that can help them respond to students' needs. Evidence from formative assessments can help teachers gain insight into students' understanding as well as provide information about the effectiveness of their own instructional strategies. By examining students' thinking as students learn mathematics, teachers can diagnose misunderstandings or faulty procedures and take advantage of opportunities to address those issues before they are ingrained and more difficult to change.

Summative assessments are also an important component of an assessment plan. Projects, chapter tests, end-of-course final examinations, graduation proficiency examinations, and state or provincial achievement tests supply information for teachers, students, parents, school officials, government officials, or other members of the public about how students, teachers, and schools are performing. At the classroom level, those assessments are used to evaluate students' understanding or the effectiveness of a particular instructional approach. At the school or district level, those assessments may have a serious effect on a student's future as well as on the school's or district's priorities. Teachers must be sensitive to the results of summative assessments, particularly of those created by bodies external to a school or district. The results of external assessments can identify programmatic strengths and weaknesses that should provide an agenda for discussion and debate among teachers and other school professionals with a view toward school or curricular improvement.

One of the most powerful forums for teacher improvement is involvement in a professional learning community.

Collaborative efforts toward school improvement can also help define an agenda for formal and informal professional development activities. Teachers have a professional responsibility to participate in group decision making to improve the art and practice of teaching (Stigler and Hiebert 1999). Attending conferences, reading professional journals, and participating in workshops and professional organizations can expand a teacher's mathematical content knowledge and enhance her or his repertoire of teaching techniques. One of the most powerful forums for teacher improvement is involvement in a professional learning community (Stigler and Hiebert 1999). In such a community, teachers select an issue on which to focus and the means by which they explore the issue and implement changes. Participation in professional learning communities is effective because it empowers teachers to set priorities and make decisions. This type of professional development also reflects the value of lifelong learning and continual improvement, as well as serves to create a model of intellectual curiosity for students.

Standard 7: Reflections on Teaching Practice

Vignette

In the vignette, a teacher assesses her teaching practice by analyzing audio-recordings of her class sessions. She focuses on the unequal participation patterns of boys and girls in her class. After reflecting on her observations and some related reading, the teacher devises a plan to further explore an issue that is of great importance to her.

7.1—*Examining Interaction Patterns in the Classroom*

Ms. Weissmann has been audiotaping her mathematics classes each day this year. She listens to as much of each tape as possible while she plans for the next day's class. In listening to herself and to the students, she begins to notice a pattern.

On the one hand, the girls are very quiet, speak softly, and say, "I don't know" at least as often as they say anything. The boys, on the other hand, are loud, and she hears herself calling them by name a lot. They participate actively in the mathematics discussions as well as in their own little games and fooling around. She begins tallying the frequency with which she calls on boys and on girls. She also begins a chart for what the boys and girls each contribute to class discussions, not just how often.

The pattern that she has noticed is not uncommon, but it is troubling to Ms. Weissmann, who has always been interested in, and relatively successful with, mathematics. She also is convinced that the classroom dynamic does not have to remain as it is.

At the same time, Ms. Weissmann gets a couple of books from the library, both centered on discourse and on women's patterns of interaction in different settings. She decides to adopt a project for herself: to improve the balance of kinds and frequency of participation among boys and girls in the class discussion. She also plans to be alert to any other patterns of interaction that may be undermining her goal of equitable opportunities to learn for all.

Concluding Thoughts
Standard 7: Reflections on Teaching Practice

Monitoring lesson effectiveness can be complemented by monitoring the overall structure, climate, and patterns of the classroom. How rich are the tasks in which students engage? Who is participating in the discourse? What types of questions are being asked, and what level of mathematical thinking is required to answer those questions? An examination of those global issues either independently or collaboratively with peers or supervisors can provide information about teaching effectiveness and identify areas for improvement. Participation in learning and professional development communities can enhance teacher knowledge and supply teachers with resources for support and improvement of teaching and learning.

Summary of the Analysis Standards

An analysis of instruction recognizes the intimate relationship among teaching, learning, and assessment. To improve their mathematics instruction, teachers must constantly analyze what they and their students are doing and how that approach is affecting what the students are learning. Using a variety of strategies, teachers must continuously monitor students' ability and inclination to analyze situations, to frame and solve problems, and to make sense of mathematical concepts and procedures. Teachers should use such information about students not only to assess how students are doing but also to appraise how well the tasks, discourse, and environment are working together to develop students' mathematical understanding and to help them adapt their instruction appropriately in response.

To improve their mathematics instruction, teachers must constantly analyze what they and their students are doing and how that approach is affecting what the students are learning.

Chapter 2

Standards for the Observation, Supervision, and Improvement of Mathematics Teaching

Overview

This section presents six Standards for the Observation, Supervision, and Improvement of Mathematics Teaching, organized under two categories:

1. The Process of Teacher Observation, Supervision, and Improvement
   A. The Continuous Improvement Cycle
   B. Teachers as Reflective Participants in the Observation, Supervision, and Improvement Process
   C. Data Sources for Observing, Supervising, and Improving Mathematics Teaching
2. The Objects of Focus in Teacher Observation, Supervision, and Improvement
   A. Teacher Knowledge and Implementation of Important Mathematics
   B. Teacher Knowledge and Implementation of Effective Learning Environments and Mathematical Discourse
   C. Teacher Knowledge and Implementation of Assessment for Students' Mathematical Understanding

Introduction

Efforts to improve the teaching of mathematics are necessarily a function of what good mathematics teaching is considered to be. Deciding which aspects of teaching need to be improved requires both information about the teaching process and a framework that suggests what is valued. The previous chapter presents a vision for teaching mathematics on the basis of *Principles and Standards for School Mathematics (Principles and Standards)* (NCTM 2000). The Standards in this chapter are intended to help teachers and administrators attain that vision by emphasizing the role that observation and supervision can play in teachers' professional growth. In keeping with the notion that assessment is a process of gathering and interpreting information, these Standards focus on how and what information should be gathered and analyzed to help teachers improve their teaching.

The assessment process described in the Standards for the Observation, Supervision, and Improvement of Mathematics Teaching can be used by a teacher engaged in a process of self-analysis for professional growth or by a teacher working in concert with colleagues, supervisors, or administrators in an effort to improve instruction. Each Standard serves as a statement about what should be observed regardless of who is doing the observing. Further, the Standards can be useful in developing teachers with a wide range of teaching experience and expertise. The Standards provide a vision that must be considered by all who teach mathematics and observe the teaching of mathematics.

The cycle of observation, supervision, and improvement is manifested in many ways. Teachers improve their teaching by reflecting on and analyzing previous lessons. Teachers improve their teaching in activity with peers, supporting one another by working together in collaborative teams to discuss best practices and to improve the quality of instruction. The vignettes in this chapter illustrate such forms of evaluation.

Teachers who seek continuous improvement in their teaching take risks by experimenting with instructional approaches that are either new to them or that they have not yet mastered.

Teachers who seek continuous improvement in their teaching take risks by experimenting with instructional approaches that are either new to them or that they have not yet mastered. The observation and supervision process should not restrict a teacher's willingness to take those risks. Teachers need freedom and support to develop professionally. For a teacher engaged in experimentation, self-analysis and collaboration with colleagues are required.

Effective mathematics teaching requires every student to learn mathematics with understanding, actively building new knowledge from experience and existing knowledge. By "every student" we mean specifically—

- students who have been denied access in any way to educational opportunities as well as those who have not;
- students who are African Americans, Hispanics/Latinos, Native Americans, Alaskan Natives, Pacific Islanders, Asian Americans, First Nations people, and other minorities, as well as those who are considered to be a part of the majority;
- students who are female as well as those who are male;
- students who are from any socioeconomic background;
- students who are native English speakers and those who are not native English speakers;
- students with disabilities and those without disabilities; and
- students who have not been successful in school and in mathematics as well as those who have been successful.

For that reason, an important consideration in evaluating the teaching of mathematics is to determine that every student has access to high-quality mathematics instruction and assessment.

Assumptions

The Standards in this section are based on the following four assumptions:

1. The goals of observing and supervising the teaching of mathematics are to improve teaching, to improve student learning, and to support professional growth.

The teacher is the essential component in high-quality mathematics education. The teacher makes decisions about curriculum and teaching methods that maximize students' learning. During the process of observation, supervision, and improvement, the focus is on students' needs and what teachers can do to better meet those needs. Professional growth opportunities are designed to extend and expand teachers' abilities to make good decisions by giving them access to a deeper understanding of mathematics, a greater knowledge about students' learning of mathematics, a greater repertoire of teaching strategies, and the ability to match their repertoire to the needs of all students.

2. All teachers can improve their teaching of mathematics.

These Standards are intended for all teachers and administrators. Whether beginning or experienced, all teachers can find some aspect of their teaching that can be improved by considering these Standards. Although experienced teachers may be more adept at self-analysis, beginning or struggling teachers can work collaboratively with administrators to reflect on the Standards and arrive at some conclusions about how their teaching can be improved.

3. What teachers learn from the observation, supervision, and improvement process is related to how observations or evaluations are conducted.

The primary emphasis of the Standards is on improvement through self-analysis and working in a collaborative and supportive environment with peers, supervisors, and administrators. When written observation reports are prepared for a teacher's personnel file, a spirit of sensitivity, mutual respect, and concern for professional growth as the primary purpose of the observation are especially important. Effective observation of mathematics teaching requires that the observer understand how to help the teacher know what students need to learn and how to challenge and support them to learn it well.

4. Because teaching is complex, the continuous improvement of teaching is complex.

Simplistic processes will not help teachers realize the vision of teaching mathematics described in *Principles and Standards*. Teaching requires the coordination of many actions, including listening, informing, stimulating, challenging, and motivating. These and other actions should be carried out in ways that are responsive to students and that take advantage of their knowledge and understanding of mathematics. A particularly sensitive issue related to the complexity of the observation of teaching is whether and how information about students' understanding of mathematics should be considered. The idea that students' progress should provide a source of information about teaching seems only reasonable. However, student performance should not be the only source

The process of collaborative analysis should highlight longitudinal improvement on a cyclical basis.

of information. It follows that any supervision process that intends to help teachers achieve the vision of teaching mathematics suggested in this document should consider the importance of teachers' working in collaborative teams to foster greater equity for student learning and improvement. The process of collaborative analysis should highlight longitudinal improvement on a cyclical basis.

The next section contains three Standards that describe the supervision process and its contribution to a teacher's professional growth. The subsequent section contains three Standards that describe some of the objects of focus during the observation, supervision, and improvement process.

The Process of Teacher Observation, Supervision, and Improvement

The process of teacher observation and supervision should generate information about teaching, and an analysis of that information should in turn lead to rich and appropriate professional growth experiences. Such professional growth should improve instruction and student learning. The Standards in this section address the supervision process and its connection with professional growth.

The complexity of teaching requires an improvement process based on information from a variety of sources and a variety of teaching situations. Teachers may demonstrate more strengths in one context than in another. A fair and valid supervision process should collect enough data from a variety of contexts to allow an accurate description of the teacher's abilities.

Improving the teaching of mathematics requires the enhancement of teacher knowledge and practice. The observation and supervision process may reveal areas of instruction that are not consistent with the desired vision of teaching mathematics, but only the teacher can take steps toward realizing that vision. Participation in high-quality professional growth experiences helps teachers become more reflective practitioners and, as a result, more attuned to the effects of their practice on students. The teacher is the essential element in the assessment process and, consequently, should be fully involved in determining which aspects of teaching should be the foci of individual and collective professional development with colleagues.

Such professional growth experiences are a natural component of the continuous improvement cycle in which teachers should engage throughout their careers:

> The process of teaching involves creating a learning community, challenging students to make sense of mathematical ideas, and supporting students' developing understanding. The teaching process involves a myriad of decisions. . . . Consequently, it is not surprising that learning to teach well is a career-long endeavor! With roots in preservice education, it is sustained throughout one's career in an ongoing process of

> learning what students understand, how they understand it, and what learning activities most effectively support meaningful and useful understanding.
>
> (NCTM 2004, p. xi)

The purpose of teacher observation, supervision, and improvement is to develop teachers' *knowledge* and understanding of teacher and student actions that lead to improved student learning. The *implementation* of a teacher's knowledge should be observable during the teacher observation, supervision, and improvement process. In addition, the observation, supervision, and improvement process should also support the *analysis* of teaching by providing opportunities for individual and collective teacher reflection.

Standard 1: The Continuous Improvement Cycle

The observation, supervision, and improvement of the teaching of mathematics should be cyclical process involving—

- the periodic collection and analysis of information about an individual's teaching of mathematics,
- professional growth based on the analysis of teaching, and
- the improvement of teaching as a consequence of professional growth.

Elaboration

The cycle of improvement should be ongoing. Observation and supervision within a collaborative framework is the vehicle that connects a teacher's current teaching with the professional growth experiences necessary to enable that teacher to improve her or his teaching of mathematics. The process begins by collecting data representative of the teacher's current practice. The collected data are then analyzed collaboratively with respect to what is valued in the teaching of mathematics, such as the vision of teaching presented in the first chapter of this volume. Aspects of instruction that are deemed consistent with what is valued should be identified, as well as those needing improvement. Although this analysis may result in a report for the teacher's personnel file, the more important outcome is the creation of a plan to help the teacher develop professionally. That plan should consist of instruction alternatives that have the potential for improving teaching as well as strategies for implementing those alternatives. Subsequent lessons are then observed and analyzed to determine whether improvement has been made; hence, the improvement process is cyclical. Each cycle should begin with the collaborative identification of a focus for improvement on the basis of previous analysis of teacher practice.

The cycle may require only a few minutes, as when a teacher thoughtfully reviews an algebra lesson taught during one period before teaching the lesson again during a later period, or it may require a year if college course work is used as a vehicle for professional growth. Involvement in inquiry-based approaches to teaching, such as collaborative or funded projects, could translate to a multiyear cycle of improvement. More commonly, a cycle may last a few weeks. For example, if a teacher is trying to increase her repertoire of assessment techniques and is interested in determining the impact of the various techniques on students' learning of and disposition to learn mathematics, she may keep track of various assessment methods and outcomes within a given grading period.

Too often the improvement process involves only a supervisor or administrator making a single observation during an academic year. Such a process is limited in at least three ways. First, annual observations are much too infrequent to provide the basis for a comprehensive plan for professional growth. Second, suggestions for improvement by a single observer can be limiting and ignore the wealth of expertise available from the teacher and the teacher's colleagues (see Standard 2). Third, teacher improvement plans based on a single source, such as a single classroom observation, are similarly unreliable and ignore other important sources of data that furnish additional information about teaching that would be useful in setting goals for professional growth (see Standard 3).

Professional growth experiences can take many forms (see the third chapter of this volume), including independent study, participation in in-service programs offered by the school, enrollment in college courses, active and collaborative discussions with colleagues, observations of colleagues, and attendance at professional meetings. Evidence of professional growth should appear in subsequent teaching and be documented in future assessments.

The major goals for the supervision of mathematics teaching should be to improve teaching, enhance professional growth, and create greater equity in students' learning experiences.

The major goals for the supervision of mathematics teaching should be to improve teaching, enhance professional growth, and create greater equity in students' learning experiences. Such an emphasis would be a significant change from present practices in many school districts, in which the goal is to provide documentation for personnel decisions or simply to comply with a requirement that an assessment report be added to every teacher's file according to some specified schedule. Although eliminating such reports from the process may not be possible or even desirable, the primary emphasis must be placed on the use of observation to furnish the basis for professional development activities aimed at improving the teaching and learning of mathematics.

Standard 1: The Continuous Improvement Cycle

Vignette

In the following vignette, a teacher identifies a problem with allocation of time in her classroom routine by examining several videotaped teaching episodes. The teacher discusses the issue with two of her colleagues, and together they propose some strategies for improvement. By informally recording data, the teacher is able to document evidence of improvement.

1.1—*The Homework Dilemma*

Three weeks before school begins, Mary Fisher examines two videotapes of her ninth-grade algebra class from the previous spring. As a seventeen-year veteran, she does a yearly review of goals and expectations before each new school year begins. The videotape helps her recall ending last year feeling disappointed in her inability to find more classroom time to devote to problem-solving activities and discussions. She realizes the need to find time to integrate student exploration in technological and non-technological environments into the required curriculum. Yet she barely has enough time to address homework questions and present new material.

Self-analysis of the videotape reveals a startling observation regarding her allocation of class time. She notices that she spends almost twenty-five minutes, nearly the first half of the period, dealing with homework. More important, she observes that many of the students are off-task or passive while she does the problems on the board.

Mary calls Delores Laco and Chris Carter, her ninth-grade algebra teammates. She asks their opinion about changing their homework-review techniques this fall. Delores tells Mary and Chris about a recent journal article that she has read during the summer. The article offers various suggestions for reviewing homework. Mary and Chris read the article, and they all agree to discuss alternatives for covering homework when they meet during the in-service days the last week in August.

Subsequently, Mary, Delores, and Chris decide to try four different methods of going over homework during the first quarter:

1. Have students keep their homework in a notebook that will be periodically reviewed.
2. Pair students to discuss their homework briefly at the beginning of the class period.
3. Give frequent short quizzes on the homework.
4. Write solutions to selected problems on a transparency, and put those solutions on the overhead at the start of the class period.

Mary discusses the proposed strategies with the students during the first week of school. In addition, she decides to start each class period with a problem-solving activity, moving homework to later in the period.

In an effort to monitor her current classroom time spent on homework, Mary keeps a daily log of the amount of time spent going over homework in class. Delores suggests that they also keep track of the part of the period in which homework review takes place. They all agree to develop a homework assignment sheet to be used by all students in algebra. In early October, Mary reviews her second-hour algebra class log for the previous week:

	M	T	W	TH	F
Time Started	9:18	9:40	9:33	9:20	No School
Total Time	12 Min.	5 Min.	14 Min.	9 Min.	

Mary shares the results with Chris and Delores. Mary is very pleased that she has been able to reduce significantly the amount of time she has been spending on homework. Yet she has not detected any drop in student performance as a result of this new approach. On the contrary, the students seem to be more attentive when discussing homework. In addition, she is pleased with her attempts to engage students in more problem-solving activities and discussions.

Closing Thoughts

Standard 1: The Continuous Improvement Cycle

Mathematics teachers must act as reflective practitioners, taking responsibility for periodic review and collection of information for improvement.

Mathematics teachers must act as reflective practitioners, taking responsibility for periodic review and collection of information for improvement. Teachers must also seek to know and understand "best practice" in the field. The reflection-and-analysis process should include ongoing and periodic discussions and observations with colleagues and supervisors to gain greater knowledge and understanding for improved teaching.

Standard 2: Teachers as Participants in the Observation, Supervision, and Improvement Process

The improvement of the teaching of mathematics should provide ongoing opportunities for teachers to—

- analyze their own teaching,
- deliberate with colleagues about their teaching, and
- confer with supervisors and administrators about their teaching.

Elaboration

The emphasis in this Standard is on the teacher's becoming a significant participant throughout the improvement and growth process. In particular, teachers should be given the opportunity and encouragement to engage in the practice of reflecting on and evaluating their own teaching and to discuss their teaching with supervisors or other administrators who have observed their teaching. When supervisors or administrators conduct observations, teachers must play a central role in providing information about their own teaching, including their goals and analyses of teaching. Teachers must see the improvement process as one that contributes to their professional growth as teachers of mathematics, thereby necessitating their participation in the process.

Supervisors and administrators should establish collegial relationships with teachers in ways that foster an atmosphere conducive to improved teaching practices for student learning. Growth is nurtured when all parties who are interested in improving the teaching of mathematics realize that improvement is an activity for all teachers, regardless of their years of experience, tenure status, or continuing certification requirements. Networks of teachers who meet to share resources, discuss student work, or tackle difficult classroom issues together enhance shared leadership and reinforce the focus on student understanding (Neumann and Simmons 2000). As an extension, teachers who engage in more formal action research can learn to view their own practice or their students' thinking in a new light.

When observation is intended to produce a report on a teacher's competence, teachers should be given opportunities to contribute their interpretation of events and to share their instructional goals and expectations of students. Further, teachers should have access to any information accumulated during the process.

Administrators should create a culture in which teachers are required to solicit colleagues' help in guiding the improvement of instruction. In fact, "[c]reating a collaborative environment [of adults] has been called the single most important factor in sustaining the effort to create a learning community" (DuFour and Eaker 1998, p. 130). Teacher teams, organized around common goals, can be effective vehicles for such collaboration. Dufour and Eaker outlined four prerequisites for embedding such teams into a school's culture:

Administrators should create a culture in which teachers are required to solicit colleagues' help in guiding the improvement of instruction.

1. Dedicated time for collaboration must be part of the school day.
2. The purpose of the collaboration must be explicit.
3. Teachers must have explicit instruction in how to be effective collaborators.
4. Teachers must accept their individual and collective responsibilities in these professional partnerships.

Because collaboration has not been the norm in much of United States and Canadian schooling, teacher teams must be prepared for working in teams, must know what is expected of their team, must be personally invested in the team's successes, and must be given the time and tools to succeed. Even in the context of self-analysis, teachers should have confidence that the administration is supportive of standards focused on change and will provide resources necessary to initiate that change.

Because of their unique role in schools, mathematics specialists or coaches should receive adequate preparation in peer coaching as well as adequate resources with which to provide teachers the help they seek.

School mathematics specialists or coaches can play a powerful role in the continuous improvement cycle, particularly if their responsibilities are limited to helping teachers enhance their practice instead of serving in an evaluative capacity. Such area experts can serve as valuable resources for helping teachers analyze their teaching in a supportive, collegial environment with the goal of mastering a greater repertoire of more effective instructional strategies. Because of their unique role in schools, mathematics specialists or coaches should receive adequate preparation in peer coaching as well as adequate resources with which to provide teachers the help they seek.

Standard 2: Teachers as Participants in the Observation, Supervision, and Improvement Process

Vignette

In the following vignette, a teacher and supervisor work together to explore the appropriate use of technological resources to enhance student understanding. The supervisor shares information from professional resources and suggests several problems for the teacher to have his students investigate using a calculator. The supervisor also seeks additional information and support for the classroom teacher.

2.1—*Using Technology to Enhance Learning*

Pete Wilder has been teaching eighth-grade mathematics at Blackhawk Middle School for the past twenty years. Although he has read that calculators can enhance students' understanding and use of numbers and operations, he has been reluctant to allow his students to use them very often. Pete is concerned that his students will lose their ability to perform computations by hand. His supervisor, Tim Jackson, has been urging Pete to explore the appropriate use of this technology to expand students' mathematical understanding, not to replace it. Tim has been disappointed that although Pete has explored potential uses of calculators in his classroom, he still appears hesitant to implement technology-rich lessons. Pete admits that most of his students use calculators only for checking computations with whole numbers or decimals.

Pete agrees to make a greater effort to incorporate calculators into his lessons if Tim will come to his class and make specific suggestions on how they can be used in a reasonable way. On the day Tim observes the class, the students are doing computational problems using the memory key on the calculator. In the course of the lesson, the students discover that the memory functions do not all work the same way on the different calculators. This variation causes Pete problems in conducting the lesson, and he is clearly not happy with the lesson. Afterward he expresses his frustration to Tim and indicates that lessons like this one are among the reasons he hesitates to use calculators with his students.

Later in the day, Pete and Tim meet to discuss the problem. Pete is quite discouraged, and this state of mind contributes to his anxiety about using calculators. Tim has two concerns. First, he wants to help Pete feel more comfortable and confident in using calculators. Second, he wants Pete to use calculators to enhance his students' learning.

Tim refers Pete to the Technology Principle in *Principles and Standards*. The Technology Principle states, "Technology is essential in teaching and learning mathematics; it influences the mathematics that is taught and enhances students' learning" (NCTM 2000, p. 24). Tim challenges Pete to determine how calculator technology can improve students' understanding of the lesson goals and outcomes.

Tim and Pete decide that they need more information on what his students know about using the calculator. Pete remembers that Juanita Criswell, another eighth-grade teacher, developed an activity sheet that she used to assess students' understanding and competence in using different calculator functions. Pete is sure that Juanita would be willing to let him use the activity sheet with his students. Pete and Tim agree that this idea is a good one. Tim also agrees to seek district funds to obtain a classroom set of calculators and to send Pete to a professional conference focused on the meaningful use of calculators and other technology in middle-grades mathematics classrooms.

Tim and Pete spend the rest of the time talking about how the calculator could be used to explore mathematics rather than just for checking computational exercises. Pete indicates that the next unit will be on statistics—finding medians and means. Tim suggests using the calculator to investigate the following types of problems:

- Suppose ten students had test scores of 68, 73, 77, 81, 84, 88, 89, 91, 94, and 95. What is the average score? Suppose the teacher made a mistake and each student should receive an additional 3 points. What will be the new average? Make a guess, and then verify using the calculator. Suppose only five of the students earned the additional 3 points. How would the average have been affected in that situation?
- If five workers each earn $38,500 a year and one of the workers receives a $5,000 raise, how much will the average salary increase? Describe how two workers could have received raises and resulted in the same average salary increase.
- If four out of six workers earn $36,000, $42,000, $37,000, and $41,000 and the average for the six workers is $40,000, how much do the fifth and sixth workers each earn?

Pete suggests that these kinds of questions are quite different from most that he has asked in the past. He likes them and indicates that he is willing to involve students in such explorations using calculators once he knows more about how well the students can use calculators. Tim also agrees to investigate what other software or resources are available to enhance students' learning of statistics.

Closing Thoughts
Standard 2: Teachers as Participants in the Observation, Supervision, and Improvement Process

Teachers must be active participants in the improvement of the teaching process. Important in this Standard is that the deliberation and discussion of teacher improvement with colleagues or supervisors are focused on improving students' understanding and their dispositions toward mathematics. In schools that reflect this Standard, all the teachers, supervisors, and administrators are committed to working together to improve their knowledge of teaching focused on improved student understanding and performance.

Standard 3: Data Sources for the Observation, Supervision, and Improvement of Mathematics Teaching

Improvement in the teaching of mathematics should be based on information from a variety of data sources including the teacher's—

- goals and expectations for students' learning and achievement;
- plans for addressing the students' achievement goals;
- lesson plans, student activities and materials, and means of assessing students' understanding of mathematics;
- analysis of multiple episodes of classroom teaching;
- analysis of classroom teaching;
- evidence of students' understanding of mathematics; and
- participation in collaborative activities with colleagues.

Elaboration

Any evaluation of teaching should be based on multiple observations and a variety of types of data—not on a single observation or a single course or type of information. A teacher's goals and expectations for students' learning and achievement should be part of the information used in the improvement process and should be discussed with the teacher prior to classroom observations. An important facet of any improvement process is that it maintain a longitudinal and cyclical orientation by considering student achievement goals previously established and those looking to the future.

Evaluation should be based on observations of the teacher teaching mathematics in a variety of contexts.

Evaluation should be based on observations of the teacher teaching mathematics in a variety of contexts—that is, at different grade levels, with a variety of students, and across mathematical topics. Such variety provides a sound basis for observing a teacher's expertise in using various teaching methods. Some visitations should be on consecutive days to understand better the continuity of classroom events. As suggested in Standard 1, those sources of information can be used either by a teacher engaged in self-assessment or by colleagues and supervisors working collegially with the teacher.

Clearly, to expect that all possible teaching situations for a given teacher can be observed would be unreasonable. Accordingly, teachers should share lesson plans, student activities and materials, projects, and student assessment techniques that they have gathered over a period of time. The opportunities to share and collaborate should be provided within the scope of a normal working day and should supplement observations of the teaching process. Teachers should use an ongoing collection of sample materials, including formative assessments, to provide information about "life in the classroom" as a basis for self-analysis and for discussion with colleagues and administrators.

The opportunities to share and collaborate should be provided within the scope of a normal working day and should supplement observations of the teaching process.

A teacher's analysis of the teaching and learning process could furnish valuable information about what the teacher is intending to accomplish and may also provide a basis for improvement. The inclusion of multiple episodes in such analysis may also provide a frame of reference for how specific lessons fall within the teacher's repertoire of instructional strategies and expectations for students.

Evidence of students' understanding of mathematics should provide a source of information about teaching, but it should not be the only source of information. Furthermore, learning should be considered with respect to every student; that is, increasing the learning of mathematics for some students at the expense of neglecting other students is not appropriate. Finally, an assessment of students' learning mathematics should be based on the full range of mathematical activity described by NCTM in *Principles and Standards* (2000). It should not be based on a narrow range of specified objectives.

Teaching is no longer a solitary activity. Teaching has become a professional activity open to collective observations, study, and improvement. As teachers openly share and study their practice of mathematics teaching with one another and with supervisors, the resulting collaboration has the potential to have a greater impact on students' performance than any single teacher can have working in isolation. One such method for collaborative planning and improvement of lessons, pioneered in elementary schools in Japan, is called Lesson Study (Stigler and Hiebert 1999).

Teaching is no longer a solitary activity. Teaching has become a professional activity open to collective observations, study, and improvement.

Lesson Study can be conducted in many ways, but the central components involve teachers working together to identify an area of student difficulty, creating a lesson that addresses students' needs, piloting the lesson in one teacher's classroom, making improvements to the lesson, and continuing the cycle of lesson delivery and improvement until the team of teachers is satisfied with the quality of the lesson for meeting the needs of students. The use of Lesson Study creates an atmosphere of professional cooperation and makes greater use of a school's or district's intellectual resources for the improvement of teaching and learning.

Many other types of collegial interaction can contribute to developing better instructional models and improving students' learning, including (a) professional development teams composed of teachers at a given school or of preservice teachers working with a university faculty member; (b) high-level reflective discussions about common mathematical or pedagogical concerns; (c) frequent supportive observation and helpful critique of classroom teachers by peers; (d) collaborative teacher research, both formal and informal; (c) and teachers teaching other teachers in professional development settings, both global, such as conferences, workshops, and seminars, and local, such as planned department meetings (Taylor 2004).

Standard 3: Data Sources for the Observation, Supervision, and Improvement of Mathematics Teaching

Vignette

In the first vignette, a supervisor discovers the benefit of observing multiple class periods. Whereas one class was more routine and uninteresting on the day she observed, the second class period demonstrated more of the teacher's creativity and ability to get students to make connections and become excited about mathematics.

3.1—*Multiple Episodes of Classroom Teaching*

Sandi Rodriguez, the principal at Westside Middle School, is visiting Ben Bedo's seventh-grade mathematics class. Before observing the class, she talked with Ben about his specific student outcomes and activities for the class. She notes that he is always well organized. However, her first observation during the third week of school left her with the impression that his teaching rarely goes beyond what is presented in the textbook; the examples and activities he used were all from the textbook. Still, she is impressed with Ben's ability to describe individual students and to diagnose individual learning problems. He seems to know his students very well—especially considering that it is only October.

Ben begins class by addressing homework questions on changing written statements to simple equations. After asking students to answer the questions, Ben introduces the new material involving writing equations based on data presented in a table using the examples in the textbook. He has the students work a few problems and then assigns problems from the textbook. The balance of the period is spent on students' completing the homework assignment as Ben tours the student work teams and helps them. He manages the class very well.

Mrs. Rodriguez makes a note for their planned after-school conference to encourage Mr. Bedo to be more focused on *student understanding*. She contemplates giving Ben high marks for organization, knowledge of the content, and general manner in conducting the class. She observes that the room is attractively arranged with interesting bulletin boards—suggesting some degree of creativity. She is considering giving him a below-average rating on his ability to challenge students to think more deeply about the problems they are solving and to make connections that encourage problem solving and communication in the classroom. She is generally unimpressed with the development of new material.

The bell rings, and students begin to pack up and leave. Mrs. Rodriguez is ready to leave as well when suddenly a rush of students comes into the room for the next class. They are quite excited. Mrs. Rodriguez decides to stay a few more minutes and see what is going on.

Maria has a branch of a poplar tree, Hector has a branch of a weeping willow tree, and Paul has a branch of an almond tree.

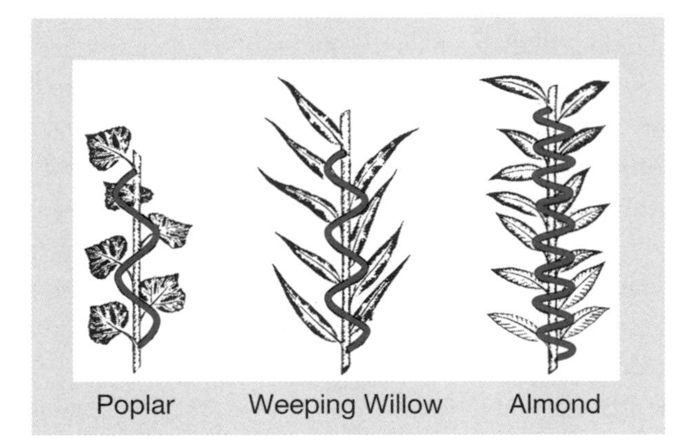

Other students have objects that represent spirals—seashells, pinecones, and sunflower heads. Mrs. Rodriguez hears one of the students proclaim, "Oh! Look at that one! Cool!" Mrs. Rodriguez decides she needs to observe this second class in addition to the first class.

The students work in small groups as class begins, and as Mr. Bedo has requested, they begin talking about the way the leaves are arranged on the branches. The students are a little confused about how any of this pursuit relates to mathematics. Mr. Bedo asks them to note how the leaves are arranged on the branches. After some discussion, the students observe that the numbers 3, 5, and 8 correspond to the arrangements of the leaves.

Mr. Bedo asks whether those numbers mean anything. Hector remembers that the numbers are part of the Fibonacci sequence, which they had studied earlier. Several students ask whether this coincidence is owing to luck or whether mathematics relates to other things in life. Mr. Bedo asks them to take out the examples of spirals that they brought to class.

Mrs. Rodriguez notes Ben's efforts to make mathematics interesting and create student excitement. What a difference her unplanned observations of Ben's second-period class will make in her evaluation of his teaching. She is looking forward to discussing the two lessons with Mr. Bedo after school and benefiting from his reflection and analysis.

Vignette

In the second vignette, a group of teachers engages in a Lesson Study experience. The teachers use a variety of resources to plan and implement their lesson. Both the immediate supervisor and the principal encourage the teachers' collaborative efforts and provide logistic and financial support as needed.

3.2—*The Power of Lesson Study*

At Martin Luther King High School, algebra teachers Darshan, Eric, David, and Liz decide to form a Lesson Study team as one method for the improvement and analysis of their practice. As a team, they select two lessons they will develop together. The first lesson will be from chapter 6, "Solving Systems of Linear Equations." The second lesson, for second semester, will be from chapter 10, "Solving Quadratic Equations Using Visual Representations and Graphing Calculator Technology."

Each team member provides sample lessons, student materials, assessment results, and previously used methods for developing understanding.

The bulk of the team's work occurs in the planning and preparation stage. Each team member provides sample lessons, student materials, assessment results, and previously used methods for developing understanding. Darshan agrees to be the team member who will teach the first classroom lesson. Eric, David, and Liz begin to collect data on student thinking, learning, engagement, and behavior when required to learn how to write and solve systems of equations. The team meets on several occasions to share and discuss their existing lessons on this topic, explaining what they thought worked and what they thought could be improved. Together, the algebra team prepares the lesson that Darshan will teach. As they plan, they try to anticipate students' responses to various aspects of the lesson and how they believe Darshan should respond. They plan for small-group discourse and the use of technology as methods of checking for student understanding.

As a team, they recognize potential barriers to strategies for differentiation in their instruction that could impede student learning. Relying on the research from Tomlinson (2005), they have supported the detracking of students out of a low-level introductory algebra course into the current college-preparatory algebra curriculum. As a team, they understand the wide range of ethnic, cultural, and linguistic diversity that now exists in their algebra classes, including the mainstreamed special education students with Individualized Educational Plans. More than ever, their Lesson Study focus must determine how they will establish clear essential outcomes, employ effective management skills for a flexible classroom, and create a safe environment for the full participation of *all* students. To that end, as a team, they brainstorm ideas that will allow the lesson to be meaningful, challenging, but not too hard. They discuss tasks that will tap into students' interest, reflect various cultures, and capitalize on different learning styles.

Darshan reveals that the day of this lesson falls on the celebration of the new year in his Indian culture. He agrees to wear a "special suit" reserved to honor special holidays. He also decides to discuss his background and culture as it relates to mathematics. Specifically, Darshan plans to explain that many important mathematical concepts originated in India, including zero, the decimal system, algebra, algorithm, square root, and cube root. Throughout the year, Darshan hopes to find ways to encourage students to identify connections between their own ethnic and cultural backgrounds and mathematical concepts.

Prior to the observation, the Lesson Study team prepares copies of the lesson plan, seating chart, and any worksheets the students will be using. On the day of the study lesson, Eric, Dave, and Liz gather to watch Darshan teach the lesson they had developed together. Liz agrees to monitor and observe the student-to-student interaction patterns, including students' responses to the cultural component of the lesson. Eric examines Darshan's use of the redirect technique during whole-group discourse, and David tracks the mathematical cohesiveness of the lesson.

After the lesson, the team meets to debrief. They share the data collected during the lesson. What is the evidence that the goals for student learning were met? Did the students connect with, and find meaning in, the lesson? Were all students able to solve the systems of equations? What part of the lesson and instruction should be considered for improvement? They all agree that the discussion of the cultural connections seemed to enhance students' interest in the lesson that day.

After their group analysis, they agree that the lesson generally went well, but they were surprised by the off-task behavior of some students. They brainstorm methods for changing that aspect of the lesson next time. Eric volunteers to teach the Lesson Study class for the second semester, admitting it is very difficult to teach in front of all peers on his algebra team. He also hopes to find a way to work his German heritage into his lesson, following Darshan's example of helping students see connections between mathematics and culture. He decides to do some research about German-born mathematician Emmy Noether as a way of drawing on his heritage and also reminding his students that serious mathematics is for women as well as men.

Closing Thoughts
Standard 3: Data Sources for the Observation, Supervision, and Improvement of Mathematics Teaching

Data should consist of lesson plans, student work samples, and multiple observation reports based on a variety of classes on consecutive days.

Supervisors and administrators need to collect ample evidence of the full scope of teachers' work. Data should consist of lesson plans, student work samples, and multiple observation reports based on a variety of classes on consecutive days. Teachers should have input into what is observed for the purpose of sharing successes, collaboratively reflecting on goals, and seeking advice for improvement. Supervisors and administrators should pay careful attention to evidence of ongoing improvement efforts and seek resources to support teachers in their efforts to enhance students' learning. Finally, supervisors and administrators should encourage teachers to work with colleagues to share materials and strategies, as well as engage in collaborative planning.

Summary of the Process of Teacher Observation, Supervision, and Improvement Standards

The process of observation and supervision should reflect the overall intent to improve instruction for the purpose of improved student understanding.

The first three Observation, Supervision, and Improvement Standards are based on the assumptions stated at the beginning of this chapter: (1) that the process of observation and supervision should reflect the overall intent to improve instruction for the purpose of improved student understanding, (2) that it should be a dynamic and continual process, and (3) that teachers should be an integral part of that process. Because of the complexity of teaching, the observation and supervision process should involve a variety of sources of information gathered in various ways.

The Standards emphasize that teachers should be encouraged and supported to engage in self-analysis and to work with colleagues on high-performing teams to improve their teaching. The focus of the teaching should be the passionate pursuit of improved student achievement results for the mathematics program. When improvement involves supervisors or administrators, their relationships with teachers should be collegial with the intent to positively affect and improve instruction.

The next three Standards focus on what should be observed during the observation and improvement process. They assume that observers, including teachers themselves, should have a framework from which to observe classroom activities. These Standards provide that framework. The framework is emphasized in the vision of teaching presented in the first chapter of this volume and is rooted in *Principles and Standards* (NCTM 2000). The Standards involve both mathematical content and processes.

The Objects of Focus in Teacher Observation, Supervision, and Improvement

The Standards in this part of the chapter are intended to help guide supervisors and administrators as they observe and assess mathematics teachers. To assess the quality of teaching and learning in a mathematics classroom, supervisors should look for evidence that the classroom is functioning in ways that optimize students' opportunities to make sense of mathematics. Because so much goes on in a classroom, an observer is helped by viewing classroom activity through a lens that focuses on the nature of the mathematical interaction between the teacher and students. At times the teacher supervision process should focus on what the teacher is doing, and at other times, it should focus on what the students are doing.

At times the teacher supervision process should focus on what the teacher is doing, and at other times, it should focus on what the students are doing.

By focusing on the teacher, assessment can determine the teacher's command of knowledge and strategies for teaching mathematics as well as whether the teacher is providing adequate engagement for students to learn.

What Is the Teacher Doing?

- Choosing "good" problems—problems that invite exploration of an important mathematical concept and allow students the chance to solidify and extend their knowledge
- Assessing students' understanding by listening to discussions and asking students to justify their responses
- Using questioning techniques to facilitate students' learning and reasoning
- Encouraging students to explore multiple solutions
- Challenging students to think more deeply about the problems they are solving and to make connections with other ideas within mathematics
- Using multiple representations to foster a variety of mathematical perspectives
- Creating a variety of opportunities, such as group work and class discussions, for students to communicate mathematically
- Modeling appropriate mathematical language and strategies and with a disposition for solving challenging mathematical problems

By focusing on the students, assessment can determine whether the teacher has provided a context and opportunity for students to be engaged in significant and appropriate activities and whether the students use a variety of representations to demonstrate their mathematical thinking.

What Are Students Doing?

- Engaging actively in the learning process
- Using existing mathematical knowledge to make sense of assigned tasks
- Making connections among mathematical concepts
- Reasoning and making conjectures about a problem
- Communicating their mathematical thinking orally and in writing
- Listening and reacting to others' thinking and solutions to problems
- Using a variety of representations, such as pictures, tables, graphs, and words, for their mathematical thinking
- Using mathematical and technological tools, such as physical materials, calculators, and computers, along with textbooks and other instructional materials
- Building new mathematical knowledge through problem solving and understanding

To a great extent, improving the teaching of mathematics depends both on a teacher's ability to determine what individual students know and how they construct mathematical ideas and on the teacher's ability to base instruction on those determinations. Teachers must be able to analyze students' understanding of both mathematical content and mathematical processes. Teachers must also be able to analyze and manage how well groups of students reason and solve problems together and communicate their mathematical ideas. Such appraisal is vital to using group work in ways that foster the development of students' mathematical power within the classroom community. The following Standards identify components of teachers' knowledge and performance that indicate an ability to help foster students' mathematical learning.

Standard 4: Teacher Knowledge and Implementation of Important Mathematics

Assessment of the effective teaching of worthwhile mathematical tasks should provide evidence that the teacher—

- demonstrates a sound knowledge of mathematical concepts and procedures;
- represents mathematics as a network of interconnected concepts and procedures;
- emphasizes connections between mathematics and other disciplines and connections with daily living;
- models and emphasizes aspects of problem solving, including building new mathematical knowledge, applying and adapting a variety of appropriate strategies to solve problems, making and investigating mathematical conjectures, and selecting and using various types of reasoning and methods of proof;

- recognizes reasoning and proof as fundamental aspects of mathematics; and
- models and emphasizes mathematical communication to help students organize and consolidate their mathematical thinking.

Elaboration

The primary emphasis in this Standard is on the teaching of mathematical content and processes. The teacher should demonstrate a deep understanding of mathematical concepts and principles, connections between concepts and procedures, connections across mathematical topics (e.g., providing geometric interpretations of probability concepts or of factoring whole numbers), and connections between mathematics and other disciplines. A teacher with a sound knowledge of mathematics can respond appropriately to students' questions, can design appropriate learning activities involving a variety of mathematical representations, and can orchestrate mathematical discourse in the classroom. Furthermore, teachers should share a perspective that mathematics is the result of human endeavor and that they should help their students become aware of the uses of mathematics that permeate modern life.

The teacher should demonstrate a deep understanding of mathematical concepts and principles.

In contrast, making frequent mathematical mistakes, using limited or inappropriate representations, or presenting mathematics as a static subject from which meaning can be derived solely from symbolic representations indicates that the teacher does not have an acceptable command of mathematics or a broad enough perspective on the nature of mathematics.

Connections should occur frequently enough to influence students' beliefs about the value of mathematics in society and its contributions to other disciplines. Regardless of the mathematics being studied, students should have opportunities to apply the mathematics they have learned to real-world situations that go beyond the usual textbook word problems. Students should see mathematics as something that pervades society and, indeed, their own lives. This Standard implies that instructional activities and lesson planning aimed at promoting students' appreciation of mathematical connections and use of multiple representations should take advantage of students' experiences and interests.

Regardless of the mathematics being studied, students should have opportunities to apply the mathematics they have learned to real-world situations that go beyond the usual textbook word problems.

Teaching mathematics from a sense-making perspective entails more than solving nonroutine problems. Mathematical reasoning must be demonstrated as an essential skill. The very essence of studying mathematics is itself an exercise in exploring, conjecturing, examining, and testing, and in building new mathematical knowledge—all aspects of problem solving. Teachers should create and present tasks that are accessible to students and extend their existing knowledge of mathematics and problem solving. Students should be given opportunities to formulate problems from given situations and create new problems by modifying the conditions of a given problem and by linking the new problem with their existing knowledge or thinking.

The very essence of studying mathematics is itself an exercise in exploring, conjecturing, examining, and testing, and in building new mathematical knowledge—all aspects of problem solving.

Standard 4: Teacher Knowledge and Implementation of Important Mathematics

Vignette

In the first vignette, a teacher assesses her students' conceptual understanding of fraction addition by having them model examples using a number-line representation. She discovers that although the students can perform the addition procedure, they lack a basic understanding of the concept. By having students build on their informal, or commonsense, understanding of the number-line model and other representations, the teacher helps students make sense of the concept of fraction addition.

4.1—*Multiple Representations for Number and Operations*

Sara Rasmussen has been teaching fifth-grade mathematics for several years. Before that, she taught mathematics to sixth graders. Although Sara is an excellent teacher, she is continually concerned about her teaching of common fractions, particularly with her students' ability to interpret fractions in a variety of contexts and to interpret various operations with fractions. This year she has made a special effort to create tasks that require the use of multiple representations of common fractions, including the number line, regions, parts of sets, decimals, and measurement. She thinks that the tasks have helped the students develop a good grasp of translating among representations—representing 3/4 as a region within a rectangle, as a point on the number line, and as a decimal, for example. She is also pleased that her students are proficient in adding fractions.

Sara decides to determine how well the students can make the connection between the *concept* of fraction and the *addition* of fractions. She asks them to add 3/4 + 1/2 and to show their reasoning using the number line. She is surprised that the students have very little sense of how to model the addition of fractions using the number line. They can mark the points 3/4 and 1/2 and the sum, $1\frac{1}{4}$, but they fail to make the connection with finding the sum by starting at the point 3/4 and moving 1/2 of a unit to the right to obtain $1\frac{1}{4}$. She spends the greater part of one period helping the students understand this connection.

Later in the day, Sara talks with another teacher about the problem to ask whether she has any suggestions for activities that could extend what she has started. Her colleague suggests a task that requires students to reason why certain procedures for adding fractions do not work. Sara decides that she will try the suggested activity.

The next day, Sara writes the following examples on the board:

$$\frac{1}{2}+\frac{3}{6}\stackrel{?}{=}\frac{4}{8} \qquad \frac{1}{3}+\frac{1}{3}\stackrel{?}{=}\frac{2}{6}$$

$$\frac{1}{5}+\frac{1}{2}\stackrel{?}{=}\frac{2}{7} \qquad \frac{0}{4}+\frac{1}{5}\stackrel{?}{=}\frac{1}{9}$$

She asks the students to copy the examples into their journals and to write a brief explanation of why they think the examples are correct or incorrect.

A few days later, Sara collects the journals and observes the following entries:

- It couldn't be right because 1/2 plus 1/2 must be more than 1/2.
- You need like terms—like 1 centimeter and 2 millimeters is not 3 centimeters. They need to be all in centimeters or all in millimeters.
- It is a good way to add fractions because it is easy.
- If you take one half a pie and one half a pie you get a whole pie, not part of a pie.
- If you start off at 1/5 and don't add anything you wouldn't go back to 1/9. You would stay put at 1/5.
- I used my calculator and used decimals. It gave me 0.7 for 0.2 + 0.5. That problem can't be right.

She is impressed with the depth of some of the students' understanding. She is pleased that they have considered the examples in light of the curious representations of fractions they have been studying. Most of the students used the representations appropriately to show why the procedure of adding numerators and denominators does not work.

The next day, Sara arranges the class into small groups and hands out the foregoing journal entries after the students agree that it is okay to share them. She asks the students to indicate whether they agree or disagree with each of the statements and why they agree or disagree. The students discuss their reactions within their group and then share their reactions with the class. Sara believes that most of the students are making excellent contributions in analyzing the statements. One of the students illustrates the problem with the first example using both the number line and a visual representation of circle regions. Sara is impressed.

Several days later, Sara listens to a student explain why you could not add 1/3 and 5/6 unless you find a common unit for thirds and sixths. She conjectures that the explanations may have been stimulated by the previous discussions about the journal entries. She believes that she has made progress in helping students develop a stronger number sense with respect to fractions on which they can build further understandings.

Vignette

In this second vignette, a teacher uses a real-life application as the context for a lesson that has typically been difficult and uninteresting for students. Although the teacher struggles with time management during his first time trying a new approach, he discovers that the use of a student-centered problem is motivating to his students.

4.2—*Real-Life Connections*

Steve Cooper has taught at North High School for ten years. He generally uses his seventh-hour planning period to write his lesson plans for the next day, but today he is meeting with the mathematics supervisor, Mrs. Johnson, in preparation for a scheduled observation. Mr. Cooper's concern about tomorrow's algebra class dominates the discussion. The topic is writing the equation of a line given the coordinates of two points on the line. Typically he has not been able to make the topic interesting to the students, and perhaps as a consequence, the students find it difficult. Mrs. Johnson suggests connecting the lesson with an activity involving statistics and using a calculator to graph a data set and determine a line of best fit. Together they plan the lesson and discuss some questions that Mr. Cooper might use during the lesson.

The next day, Mr. Cooper begins class by asking students to guess which is longer, their foot or the inside of their arm from the wrist to the elbow. The students measure and record the lengths to the nearest centimeter. Mrs. Johnson notes the improvement in student involvement and interest.

The data for the entire class, including Mr. Cooper, are recorded on the chalkboard in tabular form, providing a numerical representation of the linear function. Mr. Cooper then asks the students to graph a scatterplot of the data with arm length on the x-axis and foot length on the y-axis. Rob and Karen enter the data into a graphing calculator at the front of the room, displaying the data and the scatterplot on the overhead screen. The other students compare their graphs with the graph on the overhead screen. Mazie notes that points that appear more than once on the table appear only once on the screen.

Algebra Class

Student	Arm (cm)	Foot (cm)
1	29	29
2	23	20
3	24	23
4	23	23
5	26	25
6	21	23
7	23	23
8	24	24
9	24	24
10	25	25
11	27	26
12	22	21
13	27	24
14	22	22
15	25	24
16	23	23
17	23	23
18	26	25
19	24	26
20	24	24
21	22	23
22	25	26
23	29	25

Mr. Cooper asks the students whether they could make a prediction about the length of a foot of a person with a thirty-five-centimeter-long forearm. In a spirited discussion, the students agree that, in general, people with longer forearms have longer feet but that making a numerical estimate from these data is difficult. Tim proposes that since ten of the people have the same arm and foot lengths, thirty-five centimeters would be a reasonable guess. Mr. Cooper uses his pen to highlight the two points (22, 22) and (25, 25) on the projec-

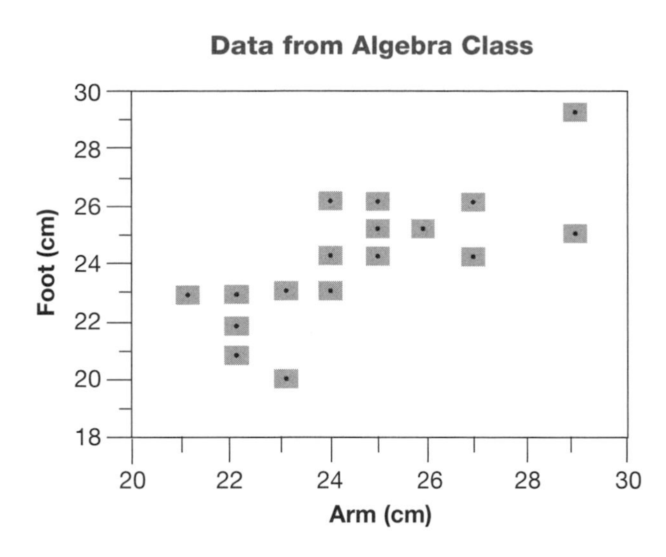

tor and then lays his pencil across those two points. The students notice that most of the points are on or close to that line. After noting that the equation of that line is $y = x$ because every point has equal x and y coordinates, Mr. Cooper helps the students understand how to find the equation of the line through (23, 23) and (26, 25). He then asks the students to pick two points on their scatterplots and follow the same technique to produce the equation of their line, the analytic representation. The students compare their equations. To conclude the lesson, Mr. Cooper has the graphing calculator generate the equation of the "best fitting" line, $y = 0.65424x + 7.9987$, and graph the line on the screen.

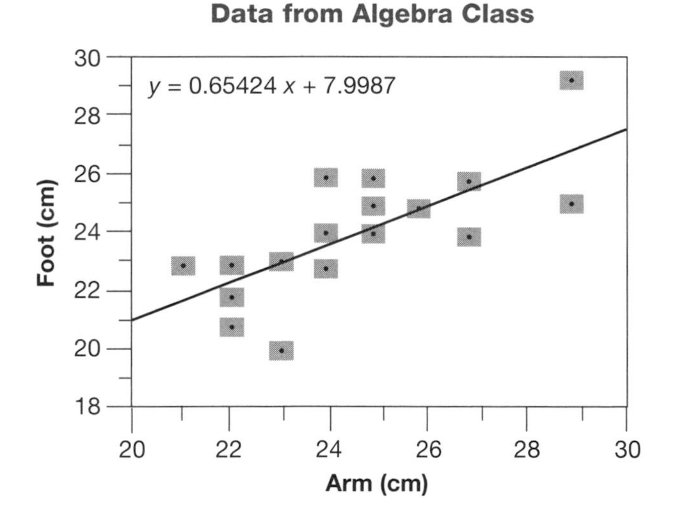

Three minutes before the bell rings Mr. Cooper asks the students to explain how they would use the calculator-generated equation to predict the length of a foot of a person with a thirty-five-centimeter forearm. He has students record their explanations on half sheets of paper. He collects those "exit" slips as a form of daily assessment to get a sense for how the students were interpreting the line-of-best-fit equation.

Later that day, during his seventh-hour planning period, Mr. Cooper and Mrs. Johnson discuss the lesson.

Mr. Cooper: This seemed to go better than the textbook approach I have used in the past.

Mrs. Johnson: The students were very engaged.

Mr. Cooper: Yes, but they still need lots of practice to get the procedure down pat.

Mrs. Johnson: Some of that practice should include situations like you used today. The students learned much more than how to write the equation of a line.

Mr. Cooper: It takes so long to do a lesson like this. I can see opportunities to explore other concepts, such as how an outlier affects the line of best fit, but I didn't have time to do anything else.

Mrs. Johnson: Student interaction in inquiry lessons like this one takes more time, but the students really seemed to understand what was happening. I think they will remember this much longer than a chalkboard explanation. You also have a great idea about exploring other statistical concepts, but perhaps computer-based dynamic statistics software would be more efficient for engaging in such explorations. Rhonda Smith has been incorporating dynamic statistics programs in her algebra classes for a while. Sometimes she makes more efficient use of her time by downloading census data and other real data from the Internet instead of having students generate the data set. You might want to get together with Rhonda and see what she suggests.

Mr. Cooper: That sounds like a good idea. I know I can use some advice about how to manage student discussion during exploration activities. I would also be more inclined to try another kind of technology tool if I knew I could get some activity ideas and management suggestions from someone else.

Mrs. Johnson: You'll find that things get easier with experience and support. Keep in mind the high level of student interest and the mathematical potential of the lesson. It was an excellent first lesson with this new material and approach.

Closing Thoughts
Standard 4: Teacher Knowledge and Implementation of Important Mathematics

Teachers should engage students in a series of tasks that involve interrelationships among mathematical concepts and procedures. The presentation of mathematical concepts and procedures means little if the content is learned in an isolated way in which connections among the various mathematical topics are neglected. Instruction should not be limited to a narrow range of outcomes, such as memorizing definitions or executing computational algorithms. Instead, instruction should incorporate a wide range of objectives as suggested in *Principles and Standards* (NCTM 2000). Further, teachers should emphasize mathematical communication with the intent of expanding students' understanding of mathematical content, its representations, and connections.

Standard 5: Teacher Knowledge and Implementation of Effective Learning Environment and Mathematical Discourse

Assessment of the student learning environment and the use of effective mathematical discourse should provide evidence that the teacher—

- engages students in tasks and representations that promote the understanding of mathematical concepts, procedures, and connections;
- engages students in mathematical discourse that extends their understanding of mathematical concepts, procedures, and connections;
- engages students in mathematical tasks that select, apply, and translate among mathematical representations to solve problems;
- engages students in mathematical discourse that communicates their mathematical thinking coherently and clearly to peers, teachers, and others, that analyzes and evaluates the mathematical thinking and strategies of others, and that uses the language of mathematics to express mathematics precisely; and
- engages students in tasks that involve problem solving, reasoning, and communication.

Elaboration

Students should learn mathematics in productive and supportive mathematics learning environments. To develop such an environment, teachers must carefully design classroom experiences so that the social culture of the classroom reflects positive values about teaching and learning mathematics. Hiebert and colleagues (1997) have identified four core features of the social culture of a classroom that promote understanding:

- Ideas and methods are valued.
- Students choose and share their methods.
- Mistakes are learning sites for everyone.
- Correctness resides in mathematical argument.

Students who learn in such an environment are expected to actively participate. To be a casual observer of others doing mathematics is not enough; everyone must take part in solving problems, describing solution strategies, and defending answers. Students also feel safe in taking on such challenges, because mistakes are expected and are seen as opportunities for learning rather than social faux pas to be avoided. In such an

Mistakes are expected and are seen as opportunities for learning rather than social faux pas to be avoided.

environment, students do not have to ask the teacher whether their answers are right to get feedback on their efforts. Reasoning, logic, and consistency measure mathematical correctness, and students are expected to know how to assess their own work and the work of others.

More specifically, problem solving, reasoning, and communication are processes that should pervade all mathematics instruction and should be modeled by teachers. Students should be engaged in mathematical tasks and discourse that require problem solving, reasoning, and communication. Consequently, administrators assessing the teaching of mathematics should determine whether teachers and students are actively involved in those processes. The ability to represent mathematics in a variety of ways takes place over time and hence should be a continuing focus of instruction and planning. It follows that assessing the existence of those processes in the teaching of mathematics must similarly take place over time.

Teachers should engage students in mathematical discourse about problem solving. Such discourse includes discussing different solutions and solution strategies for a given problem, how solutions can be extended and generalized, and that different kinds of problems can be created from a given situation. All students should be made to feel that they have something to contribute to the discussion of a problem. Questions and paraphrasing should be primary tools of the teacher as students readily apply, analyze, synthesize, and evaluate mathematical ideas individually and collectively. Assessment should focus on the quality of the teacher's questions and the richness of the class discussions. The classroom climate should give visible evidence that students are engaged in sophisticated mathematical thinking.

The goal of emphasizing reasoning in the teaching of mathematics is to empower students to reach conclusions and justify statements on their own and with peers rather than rely solely on the authority of a teacher or textbook.

Teaching mathematics as an exercise in reasoning should also be commonplace in the classroom. Students should have frequent opportunities to engage in mathematical discussions in which reasoning is valued. Students should be encouraged to explain their reasoning process for reaching a given conclusion or to justify why their particular approach to a problem is appropriate. The goal of emphasizing reasoning in the teaching of mathematics is to empower students to reach conclusions and justify statements on their own and with peers rather than rely solely on the authority of a teacher or textbook.

Students should be given the opportunity to work in groups under the watchful eye of the teacher. The teacher should be touring the classroom, observing the student work, and engaging students in communication of their mathematical thinking.

Mathematical communication can also occur when a student explains an algorithm for solving equations, when a student presents a unique method for solving a problem, when a student constructs and explains a graphical representation of real-world phenomena, or when a student offers a conjecture or proof about geometric figures. A teacher should monitor students' use of mathematical language to help develop their ability to communicate mathematics, for example, by asking students whether they agree with another student's explanation or by having students provide various representations of mathematical ideas or real-world phenomena. The emphasis must be on all students' communicating mathematics, not just on the more vocal students' doing so. For teachers to maximize communication with and among students, they should minimize the amount of time they themselves dominate classroom discussions.

Standard 5: Teacher Knowledge and Implementation of Effective Learning Environment and Mathematical Discourse

Vignette

In the first vignette, a teacher engages her kindergarten students in a dialogue about patterns. The teacher reflects on how her actions promoted discourse and student involvement. She also considers what she can do to modify and improve her activity and her facilitation of student reasoning.

5.1—*Student Involvement in Classroom Discourse*

Pat Kowalczyk's kindergarten class enjoys activities that involve continuing patterns that have been started using blocks, beads, themselves, and other items. Today, Mrs. K, as the children call her, plans to have her class construct patterns using their names. She thinks that this activity will extend the work she has been doing to encourage them to reason and communicate about mathematics with one another. She has prepared a paper with a 5×5 grid of two-centimeter squares for each student.

At their tables, the students fill out the grid, using one square for each letter of their name. When they finish writing their names the first time, they start over and continue until each of the twenty-five squares contains a letter.

K	E	N	T	K
E	N	T	K	E
N	T	K	E	N
T	K	E	N	T
K	E	N	T	K

Mrs. K.: Select your favorite crayon, and color in the squares that contain the first letter of your name.

Mrs. K walks around the room observing and listening to the students as they work. When Susan wants to know if she should color both the S's in her name, Mrs. K responds with a question, "Are they both the first letter of your name?" Susan thinks for a moment and then says, "No, only this one is," and she colors only the first S in

Susan. Mrs. K makes a mental note that Susan seems confident in her decisions and does not seek additional confirmation from her. As she continues to walk around, Mrs. K observes that some children seem to understand the activity and work independently, some are actively conferring with others, and some are waiting for her to help them. She muses, not for the first time, about what more she could do to foster greater self-reliance by her students.

When the students complete their grids, Mrs. K asks the class whether they can predict who has the same patterns of colored-in squares on their grids. She tries to phrase the question so as to encourage all the students to reason and to communicate their ideas. She asks them to raise their hands and not "call out" a response. She notices that she is improving in her ability to construct good questions on the spot.

The students quickly guess that the two Jennifers in the class should have the same pattern. Mrs. K asks several students to explain how they can be sure of their guess without even checking the girls' grids. When she calls on Marcus and he says, "Cause they have the same name, so their papers gotta be the same, too," she is really pleased. Calling on him more often seems to be paying off.

Searching for the next good question, Mrs. K challenges the students to find similar patterns where the students do not have the same first name. After some checking around, the students find that Kent's and Kyle's grids have the same pattern. Kent raises his hand and waits to be called on by Mrs. K.

Kent: Maybe names that begin with the same letter look the same.

Mrs. K: Is there anyone else whose name begins with the letter K? (Katrina, Kathy, and Kevin all jump up, waving their hands.)

Katrina: But my grid is different from Kent's and Kyle's.

Kathy: But mine is the same as Kevin's.

Mrs. K: Does this fit the rule that the names that begin with the same letter give the same pattern?

Students (in unison): No!

Mrs. K looks around, trying to decide on whom to call and tries to remember who has not spoken much today. She remembers that Nikki has not said anything today, although she did complete her grid quickly.

Mrs. K: How can we change our rule so that it works? Nikki, what do you think?

Nikki: Well, I think it will work if they have the same number of letters and if their name begins with the same letter.

Laura (excitedly): Mine matches Kathy's, but our first letters are different.

Mrs. K: Let's check it out. (She holds them up to the window, one on top of the other.) Hey, it looks like they do match!

At this point, Dave, Jane, and Jose put their patterns by Kyle's and Kent's and are surprised that the patterns match. They don't know how to express their finding. Mrs. K is a little surprised that they have difficulty explaining their finding. Judy says that it has something to do with the length of the name. Short names seem to match short names but not long names. Finally, Stanley says that the names with the same number of letters will match. Some of the other students question whether he is right. After examining many other examples, they conclude that he is correct.

After school, Mrs. K reflects on the lesson. She writes a few notes in her journal about Marcus, Nikki, and several other students. She also writes down the task so that she can remember it for the future and indicates that she thinks it could be used profitably again. She is impressed with the students' ability to reason. She thinks that letting the students use different-colored crayons to color in the grids may have distracted them from the lesson's primary objective. She makes a note to let students pick only one color the next time she uses this activity. Although she thinks she is getting better at formulating good questions, she also thinks that she needs to find more ways to encourage students to communicate their ideas with one another and to build on one another's reasoning during the whole-class discourse.

Vignette

In the second vignette, a teacher is observed as he guides students through a hands-on, technology-based investigation of the relationship between a quadratic function and its graph. The task and the cooperative setting provide a basis for student conjecture and discussion, as well as assessment of students' understanding. After the lesson, the supervisor and the teacher reflect on the quality of the classroom discourse and opportunities for extending the lesson.

5.2—*Making Mathematical Conjectures*

Art Heyen has been reading various articles in the *Mathematics Teacher* about the importance of emphasizing mathematical processes when teaching mathematics. He decides to make a concerted effort this year to incorporate those ideas into his teaching. At the beginning of the year, he meets with Diane Rowan, an experienced mathematics teacher, to discuss how his teaching could become more process oriented. Diane suggests that he start with a few selected topics to "get the feel of it" and then work from there. Diane offers suggestions for a lesson on graphing parabolas that Art could use later in the year. The lesson emphasizes the equation

$$y = ax^2 + bx + c$$

and examines the effect on the graph when different values of *a, b,* and *c* are used.

In November, Art is ready to teach the lesson on graphing parabolas. He invites Diane to observe the lesson and make suggestions. He indicates that he has had moderate success with other lessons in which mathematical processes have been emphasized. He complains somewhat about the length of time required to find good materials, but notices that the students seem more interested in mathematics this year than in either of the previous two years that he has been teaching.

Art typically teaches the lesson on graphing parabolas by modeling several graphs and helping the students locate the vertex and several other points, which they then plot. After several demonstrations, he assigns practice problems. This year, he will teach the topic with a greater emphasis on conceptual development.

Art begins the lesson by asking the students to write down three statements or words that they associate with the equation

$$y = ax^2 + bx + c.$$

Some of the phrases are *quadratic equation, parabola,* and *horseshoe shaped.* One student mentions that if $x = 0$, then $y = c$. Art asks what this outcome means, but the student is not sure. Diane thinks that Art might have spent more time helping the student reason through his conclusion.

Using a document he has previously created and saved, Art projects an image of his computer screen onto the front board. The projected page shows the graph of $y = ax^2 + bx + c$, with sliders or dynamic value indicators for the parameters, *a, b,* and *c*.

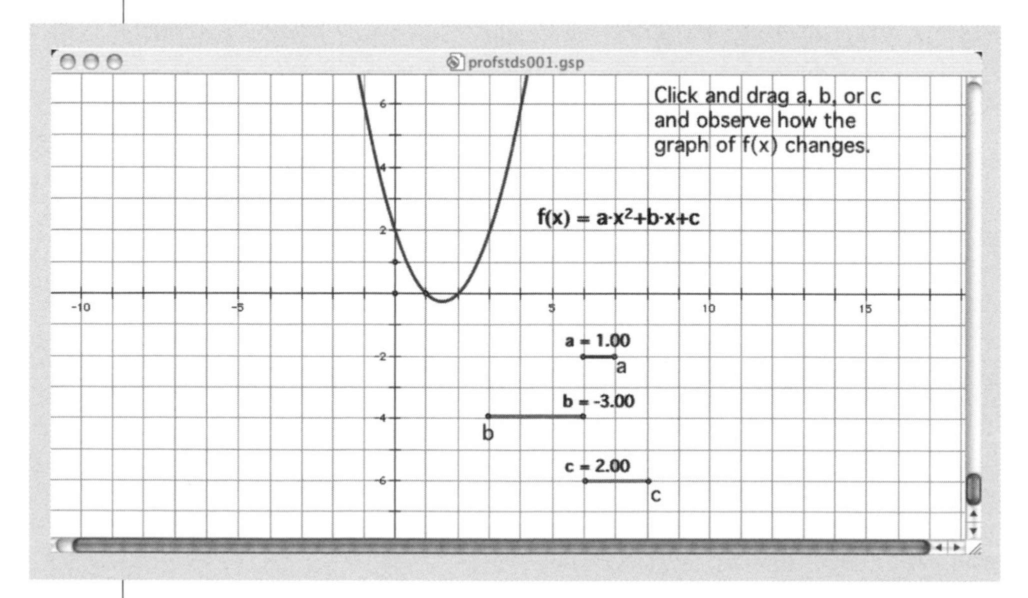

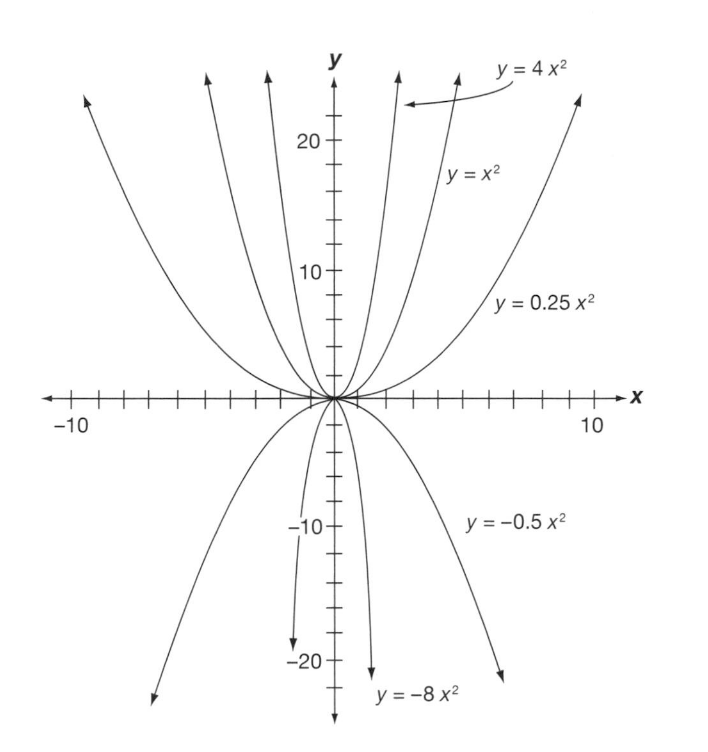

After a quick demonstration of how to drag the sliders, Art resets the sliders so that $a = 1$, $b = 0$, and $c = 0$. He invites one student at a time to come to the front and drag one of the sliders to change the value of a. Art asks students to carefully observe how the values and signs of a affect the shape, position, and orientation of the graph when the values of b and c remain constant.

Art provides graph paper for each of the students so that they can sketch some of the "interesting" variations of the equation in their notes. In addition to the basic graph, $y = x^2$, Art asks students to sketch the graphs of $y = 4x^2$, $y = \frac{1}{4}x^2$, $y = -\frac{1}{2}x^2$, and $y = -8x^2$.

The students work in teams of four to use the sketches to answer the following questions on the worksheet that Art has passed out.

Use your sketches on sheet 1 to answer the following questions:

1. What property is common to all the graphs?
2. Under what condition does the graph of $y = ax^2$ open upward? Downward?
3. As the absolute value of a increases, what happens to the graph of $y = ax^2$?

As the students work on the questions, Art and Diane walk around the room and check the students' progress. Diane notices that students have answered the first question in a variety of ways. Some students wrote that all graphs have a line of symmetry. Others noticed that the graphs are all the same sort of U-shaped curve. Still others commented that the graphs they sketched all had the point (0, 0) as their starting point. Both Art and Diane are impressed with the students' observations.

Art next invites students to vary the values of a and c to observe the characteristics of graphs of equations having the general form $y = ax^2 + c$. Students who are not at the board shout advice to the student who is dragging the sliders. Art and Dianne are impressed with the students' enthusiasm. Art directs the students to sketch the graphs of the following equations on a single set of axes in their notes.

$$y = 3x^2 + 10$$

$$y = 3x^2 + 5$$

$$y = 3x^2 - 5$$

$$y = 3x^2 - 10$$

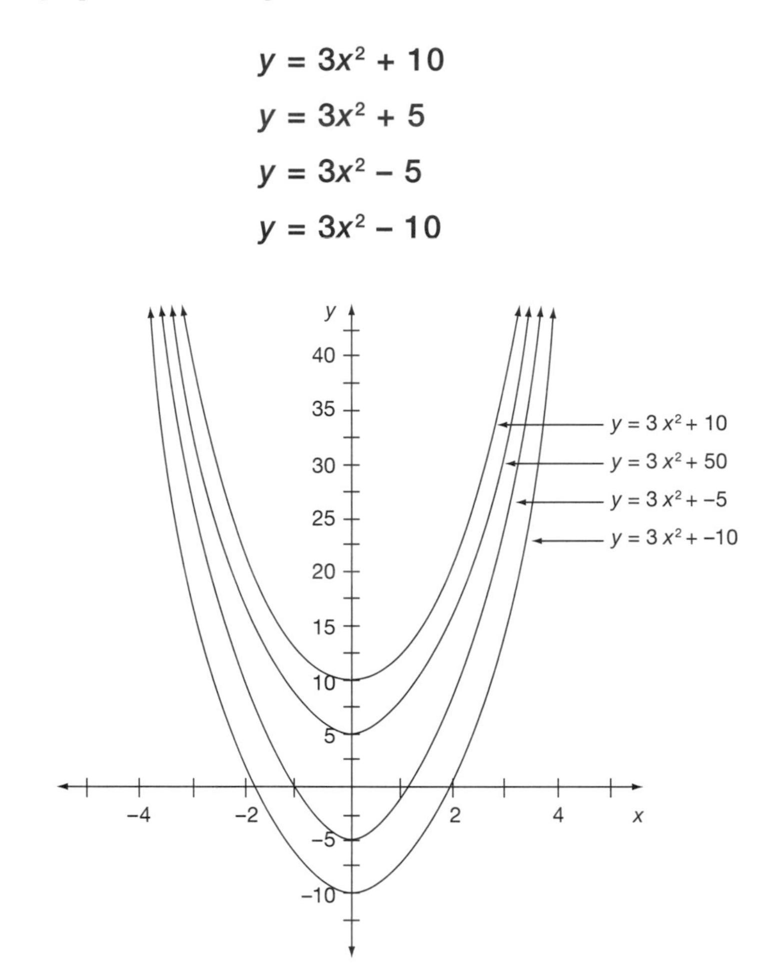

When they complete the sketches, the students are instructed to address the following questions in their teams:

1. How does the value of c affect the graph of $y = 3x^2 + c$?
2. At what point does the graph of $y = 3x^2 - 4$ intersect the y-axis?

On the basis of the graphs they have sketched, the student teams are then asked to perform the following task:

> Without using the calculator, indicate whether the graph of each of the following equations will open upward or downward; whether the graph will be relatively narrow or wide; and the point where the graph intersects the y-axis. Then sketch the graph on a sheet of graph paper.

$$y = \frac{1}{4}x^2 - 4$$

$$y = 5x^2 + 4$$

$$y = -x^2 + 7$$

$$y = \frac{1}{3}x^2 - 6$$

To conclude the lesson, Art asks the students to write one or more summary statements about what they have learned about the graphs of quadratic equations. He collects the students' papers and assigns additional equations for them to explore and sketch.

After school, Art meets briefly with Diane to discuss the lesson. He mentions that he wanted to cover more material—particularly the relationship between the value of the discriminant and the number of x-intercepts. He realizes that it would be easier to just tell the students what he wants them to know, but he was very pleased with their ability to communicate mathematics in their written statements and to use inductive reasoning to figure out what the sketches of the last set of graphs will look like.

Diane concurs. She was particularly pleased with Art's repeated efforts to encourage the students to write statements about what they had discovered and to discuss their findings with other students on their team. She suggests that next time, he might begin the lesson by asking the students what the graph of the equation $y = 1.5x^2 - 3x + 4.2$ (that is, an equation that the students would not normally encounter) would look like. After students' conjectures are recorded, the lesson could be developed, following which the equation could be revisited. The students could use their knowledge developed in the lesson to determine its graph.

Art likes the suggestion. He sees the approach as fitting in with his intention of helping students reason mathematically. Diane thanks Art for inviting her into his classroom. She compliments him on his efforts to improve his students' ability to use mathematical processes and to be actively engaged in the classroom dialogue.

Closing Thoughts

Standard 5: Teacher Knowledge and Implementation of Effective Learning Environment and Mathematical Discourse

Teachers demonstrate knowledge of an effective learning environment when they actively plan for and seek evidence that students are using inductive reasoning, proportional reasoning, and spatial reasoning and are constructing arguments. Assessing whether mathematics is being represented as a process of reasoning should focus on whether the teacher demonstrates the pervasiveness of mathematical reasoning throughout all areas of mathematics and whether the teacher requires students to use various reasoning processes.

This Standard suggests that mathematics is learned in a social context, one in which discussing ideas is valued.

Communication is the vehicle by which teachers and students can appreciate mathematics as the processes of problem solving and reasoning. Communication is also important in and of itself, because students must learn to describe phenomena through various written, verbal, and visual forms. The notion of communication emphasized in this Standard cannot be fully realized in a lecture-oriented lesson or when students' responses are limited to short answers to lower-order questions. This Standard suggests that mathematics is learned in a social context, one in which discussing ideas is valued. Classrooms should be characterized by conversations about mathematics among students and between students and the teacher as they conjecture about various mathematical concepts, skills, or procedures. Students must have opportunities for on-task, small-group discourse as the teacher observes, guides, and supports them in demonstrating an understanding of the lesson objectives.

Standard 6: Assessment for Student Understanding of Mathematics

Observing the means by which a teacher assesses students' understanding of mathematics should provide evidence that the teacher—

- uses a variety of assessment methods to determine students' understanding of mathematics as well as their ability to apply mathematics in complex and new situations;
- matches assessment methods with the students' developmental level, their mathematical maturity, and other aspects of individual diversity;
- aligns assessment methods with what is taught and how it is taught; and
- uses daily assessment practices to guide instructional practice and to make informed decisions about the learning needs of students.

Elaboration

The process of assessing teaching should determine whether and how the teacher uses evidence of students' understanding of mathematics in making instructional decisions. The assessment of students' understanding of mathematics should include methods used on a daily basis as well as those used on a less frequent basis. The latter methods include evaluating journals, notebooks, essays, and oral reports; evaluating students' homework, quizzes, and test papers; evaluating classroom discussions, including attention to students' mathematical problem-solving, communication, and reasoning processes; and evaluating group work, clinical interviews, and results of performance tests administered individually or in small groups. Such a variety of student assessment techniques reflects sensitivity to the developmental level, maturity, and cultural diversity of the students and should provide a sound basis for creating mathematical tasks and directing mathematical communication.

The Assessment Principle of *Principles and Standards* (NCTM 2000) underscores the role of assessment in supporting the learning of important mathematics. In particular, assessment is an ongoing process that feeds back information to students about their own understanding. Assessment also informs teachers about the effects of their instructional decisions and provides evidence to guide their planning. To yield such evidence, assessment tasks must be carefully designed to focus on a variety of aspects of mathematical proficiency, including understanding concepts, computing efficiently, appropriately applying concepts, reasoning in new mathematical situations, and engaging in mathematics so as to make sense of it (National Research Council 2002).

Assessment tasks must be carefully designed to focus on a variety of aspects of mathematical proficiency, including understanding concepts, computing efficiently, appropriately applying concepts, reasoning in new mathematical situations, and engaging in mathematics so as to make sense of it.

Methods of assessing students should be aligned with instructional and assessment goals. For example, if calculators or other technological tools are used throughout the instructional program, then they should be allowed in some testing situations, as well. When assessing the application of concepts or mathematical reasoning, calculators or other tools may be useful as exploratory or computational tools. However, when assessing procedural performance or computational skills, having students complete tasks by hand is more appropriate to ensure that they have mastered those skills.

Instructional activities should be based on information obtained from assessing students' mathematical proficiency. A teacher ought to be able to determine from an analysis of evidence, for example, why a student cannot apply a particular algorithm with a reasonable degree of accuracy. Does the student lack a conceptual basis for the algorithm? Is the student confused about the sequence of steps to be followed? Does the student have a sense of when to apply the algorithm, or does he or she apply the algorithm in inappropriate contexts?

Instructional activities should be based on information obtained from assessing students' mathematical proficiency.

The oft-used phrase "Are there any questions?" cannot reliably be used to determine whether students understand a concept. However, if a student claims that a parallelogram is a quadrilateral with two sides parallel and two sides congruent and the teacher asks other students whether they can provide counterexamples to that definition, then the teacher is engaging students in tasks to check understanding. Another assessment technique consists of asking students to work in teams to react to and evaluate one another's processes for solving a problem. As the students engage in guided practice, the teacher tours the room to collect assessment information about additional instruction necessary to remediate observed deficiencies.

Standard 6: Assessment for Student Understanding of Mathematics

Vignette

In the first vignette, a supervisor encourages a teacher to expand his repertoire of assessment techniques and to better align assessment with his instructional practices. As a result of their conversation, the teacher interviews two students and gains better insight into those students' understanding of place value.

6.1—*Assessing Student Understanding*

The elementary school principal, Barbara Moore, has been very impressed with the way second-year teacher Ed Dudley conducts his first-grade class and, in particular, with the way that he makes extensive use of manipulatives when teaching mathematics.

She notices, however, that when he evaluates students' progress, he relies on paper-and-pencil tests that appear to emphasize computational outcomes. When she talks with Mr. Dudley about this discrepancy, he indicates that written tests just seemed like a reasonable way to evaluate students—a very efficient method. He indicates, however, that he is willing to try different methods of assessing students' understanding of mathematics. Ms. Moore offers several suggestions.

A week later, Ms. Moore drops by Mr. Dudley's class to see whether he has had an opportunity to try any of the suggestions. She is pleased to note that in one part of the classroom, Mr. Dudley is interviewing students, while in another part of the room, pairs of students are playing a numeration game with cubes that can be linked and a spinner that determines the number of cubes each student receives. The students are to link the cubes whenever they have acquired a group of ten cubes. The students take turns spinning until each has had five turns. After the five spins, they are to write the number of cubes on a sheet of paper. Each student checks to see whether the other student has written the correct number of cubes. Ms. Moore is quite impressed with the game. She plans to suggest to Mr. Dudley that the activity could provide him with an excellent means of assessing students' understanding of place value.

Ms. Moore decides to observe one of the interviews that Mr. Dudley is conducting. She observes the following exchanges:

> Mr. Dudley gives two students, Jo and Annette, a different number of counting sticks. Each student is to bundle her sticks into groups of ten.
>
> *Jo:* I have 3 tens and 4 ones.
>
> *Mr. D:* Is that the same as thirty-four? (Jo hesitates. Mr. Dudley observes that she seems unsure whether the representation of 3 tens and 4 ones also represents thirty-four.)
>
> *Annette:* I think they are the same.
>
> *Jo:* I don't know. Let me count them. (Jo unbundles the sticks and begins counting by ones.)
>
> *Annette:* I have 4 tens and 2 ones. That's forty-two.
>
> *Mr. D:* How do you know?
>
> *Annette:* Look. (Points to bundles of ten.) Ten, twenty, thirty, forty, now (pointing to single sticks) forty-one, forty-two. See!
>
> *Mr. D:* That's very good, Annette. Let's see if we can help Jo. Jo, how are you coming?
>
> *Jo:* I counted and got thirty-four. They must be the same.

After Mr. Dudley finishes his interviews, he discusses the lesson with Ms. Moore. She compliments him on assessing students' understanding in much the same way he teaches mathematics. He indicates that the interviews did take some extra time but that generally they took less time than he had imagined. He is very pleased with how much he learned about each student's thinking about place value during the interviews. Ms. Moore helps Mr. Dudley develop a chart to make his assessment more systematic.

Place Value and Counting

Student	Date	Observations
Jo	Jan 21	Accurately counted by ones and bundled them into tens. Not sure what number was represented, however. Needs more work on recognizing number when given representation.
Annette	Jan 21	Appears confident. Has good command of translating between written number and representation using sticks.

Mr. Dudley appreciates the value of this chart for organizing assessment information. He comments that it will come in handy for longitudinal assessments of his students as well as provide particulars that he can point to during parent-teacher conferences.

Ms. Moore asks Mr. Dudley to compare the counting strategies of Jo versus Annette. She points out that the game serves as an excellent vehicle for assessing students' understanding but that it also lends insights that could guide his future teaching of place value. Mr. Dudley indicates that he needs to provide Jo with more opportunities to skip count by twos, fives, tens, and hundreds to help develop her understanding of place value. He also realizes that counting by multiples lays the groundwork for important connections between addition and multiplication.

Vignette

In the next vignette, a team of teachers works together to develop a plan to address the high rate of students repeating first-year algebra. After sharing frustrations and brainstorming ideas, they devise a plan that incorporates a variety of assessment techniques and promotes learning.

6.2—*Using a Variety of Assessment methods*

In the mathematics department at West High School, "teacher teams" have the responsibility for developing course goals and monitoring curricular changes. The "algebra team" of Alicia Washington, John Nystrom, and Katie Cusciaro teaches all ten of the first-year algebra classes at West. Through a paid summer curriculum project, their principal, Simone Richardson, has asked them to develop a plan to reduce the high number of students repeating first-year algebra. The teachers are concerned as well, because they know that a high proportion, 43 percent, of the students have been receiving Ds and Fs in algebra.

Ms. Richardson asks the team to pay particular attention to the diverse backgrounds of the students at West and the need to review fundamental geometry concepts because of the state-required test-content expectations.

In July, the three teachers meet to discuss the problem. They agree that many factors contribute to poor student performance and their high rate of algebra repeaters. They decide to focus on their teaching and assessing techniques as a first step to improving the situation. With respect to assessment, the teachers share their tests, quizzes, and the means by which they assign grades. Alicia indicates that many students do not complete her tests—they skip many items. She suspects that since English is not the first language for many of the students, they have difficulty reading some of the questions. She says that about 80 percent of her grades are based on tests and quizzes and that usually the students do not do well on tests and quizzes. Yet she thinks that they demonstrate a reasonable understanding during class discussions.

Katie is frustrated as well. She tries to create items in which the students are required to "explain" or "draw and label." She shares the following items:

1. Draw and label altitude $\overline{AG}$ for $\triangle$ ABC. Explain why it is an altitude.

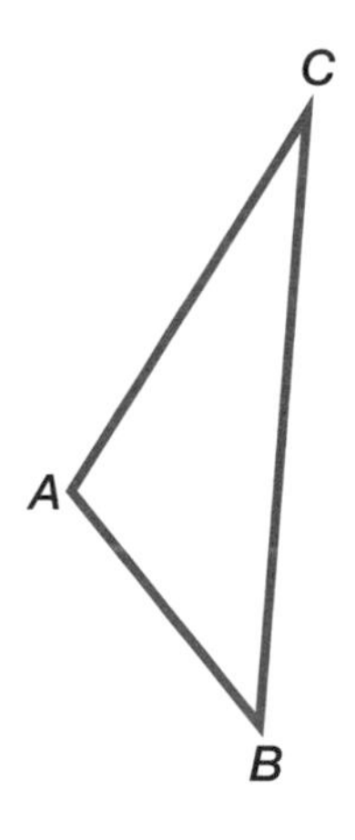

2. A student claims that x^2 is always larger than x. Is she correct? Explain your reasoning.
3. Draw and label a rectangle whose area is $x^2 + x$ cm^2.

She states that her students have a great deal of difficulty in writing mathematics—perhaps because of the language problem. Katie indicates that she has allowed students to work together when solving problems but not when taking tests or quizzes.

John indicates that the absentee rate in his class is very high—sometimes approaching 35 percent. How can he teach them if they do not come to class? He questions whether Katie is expecting too much when she wants her students to "explain" and "draw and label." After all, many students are not fluent with basic computations. He claims that he keeps it simple by sticking strictly with the tests in the book, administering one test every Friday, with make-ups on Monday. If students do not show up for the make-up tests, the low grades cannot be his fault, he argues.

The teachers continue to discuss the problem. They realize that although some things are beyond their control, they can control some things, for example, how they assess their students. They also realize that the cultural diversity of the students may require them to adjust how they have been testing and grading students. They decide on the following means of evaluating students:

1. *Journals*. Every student will keep a journal. The journal will count the same as a test grade. The daily entries will include examples worked out in class and various methods presented in class for solving problems. In addition, they will focus on the following items:
    a. What they learned that day
    b. What they did in class that helped them learn
    c. Why they think it is important to learn it
    d. How they felt about the class that day

    The team thought that those questions would also keep them on their toes when preparing lessons. For example, question "c" will serve as a constant reminder to give reasons why a topic is important to learn.
2. *Class discussions.* Greater emphasis will be given to class discussion in evaluating students, thereby encouraging students to attend class. Students will be given more responsibility to present solutions and to explain procedures during class discussions.
3. *Quizzes.* All quizzes will be taken in pairs; students will be able to discuss their solutions with their partners. This format will facilitate communication and help the students feel less tension when taking a quiz.
4. *Tests.* All tests, including makeup tests, will be shared among algebra team members. The tests will maintain an emphasis on "explaining" and will include some open-ended items as well, but they will give the students greater latitude in responding. The team will also make the tests shorter so that students will have more time to respond.

The three teachers agree to reevaluate the effect of the changes in January. Simone is impressed with what the teachers have done. She says that she will be interested in their student results during first semester. She reminds them that the ultimate goal is to increase learning and to encourage more students to be successful in doing mathematics and to continue their study of mathematics. The high failure rate is counterproductive to achieving that goal. The teachers share ideas and make modest revisions. They hope that by this time next year, they can point to the program as a model of success in increasing achievement in and disposition to do mathematics. Simone is hopeful that the program will also help reduce the absentee rate among students.

Closing Thoughts
Standard 6: Assessment for Student Understanding of Mathematics

The teacher must respect students' ideas by listening to them and incorporating those ideas into the class discussion.

Engaging students in extensive mathematical discussions using whole-group and small-group discourse, and encouraging them to reason mathematically can promote an inquiry-oriented classroom. The teacher must respect students' ideas by listening to them and incorporating those ideas into the class discussion. For their part, students should demonstrate a willingness to propose hypotheses, to support their own hypotheses, and to support or challenge hypotheses set forth by the teacher or other students. As the teacher models encouragement and support for students and respects and accepts their ideas, so should students learn to support and respect one another and to work collaboratively and actively to solve problems and to validate proposed solutions.

Summary of the Objects of Focus in Observation, Supervision, and Improvement Standards

Students must be engaged in an environment in which the processes of *doing* mathematics are continually emphasized.

Taken together, the three Standards in this section on the objects of teacher observation, supervision, and improvement present a vision for the improvement of the teaching of mathematics. These Standards emphasize the importance of significant mathematics when evaluating the teaching of mathematics. Students develop mathematical proficiency through encountering significant mathematics. To achieve that outcome, students must be engaged in an environment in which the processes of *doing* mathematics are continually emphasized.

Just such an environment can be achieved by teachers who present stimulating tasks and create evidence of an atmosphere in which problem solving, reasoning, multiple representations, and communication are valued and promoted. Such an environment requires teachers to develop lessons that motivate and foster meaning for various levels of student learners. Further, the message that teachers send students should not be limited to instruction alone; it must also include what and how mathematical learning is assessed. Through assessment we communicate to our students the mathematical outcomes that we value and that are a focus for improvement.

A consistent message throughout the Standards for the Observation, Supervision, and Improvement of Teaching is the importance of teachers' being reflective about their teaching and working with colleagues and supervisors to improve their teaching practice for student learning. Although the Standards in this section can provide a focus for improvement, such improvement is more likely to occur when teachers have the support to engage in professional development and to collaborate with one another. Professional learning opportunities must span a teacher's professional life if continuous growth and improvement are the desired goals. In the next section, Standards for the Education and Continued Professional Growth of Teachers of Mathematics are presented and discussed.

Chapter 3

Standards for the Education and Continued Professional Growth of Teachers of Mathematics

Overview

This section presents five Standards for the education and professional development of teachers of mathematics. These Standards address the following major concerns:

1. Teachers' mathematical learning experiences
2. Knowledge of mathematical content
3. Knowledge of students as learners of mathematics
4. Knowledge of mathematical pedagogy
5. Participation in career-long professional growth

Introduction

Teaching mathematics is a complex endeavor. It requires knowledge of mathematics, of students as learners of the subject, and of instructional strategies (National Research Council 2001). It calls for an understanding of the impact that socioeconomic background, cultural heritage, attitudes and beliefs, and political climate have on the learning environment (Secada 2000b). Above all, teaching mathematics calls for sensitivity and responsiveness to learners, dedication to the goal of high-quality mathematical learning opportunities for all students, and commitment to lifelong professional growth.

The Standards for the Education and Continued Professional Growth of Teachers of Mathematics address the needs of preservice as well as in-service teachers of mathematics at the pre-K–12 grade levels. These Standards apply to introductory programs that prepare teachers of mathematics; programs that provide advanced study for teachers of mathematics; and various continuing education seminars, workshops, and inquiry-oriented analyses of teaching and learning as well as other learning experiences in which teachers of mathematics participate throughout their careers.

In the early stages of their careers, preservice teachers of mathematics are focused on developing their abilities to teach mathematics. This development includes acquiring specialized knowledge of mathematics for the purpose of teaching, because teachers "must understand mathematics in ways that allow them to explain and unpack ideas in ways not needed in ordinary adult life" (National Research Council 2001, p. 371). Prospective teachers also need knowledge of students and how they develop as mathematical thinkers, and knowledge of instructional strategies, resources, and techniques for creating productive learning environments for their students. Finally, teacher preparation involves

During their preservice period, prospective teachers can explore, analyze, and question new ideas with support and encouragement.

Initial teaching assignments and support structures play a significant role in shaping beginning teachers' views of the profession as well as their commitment to it.

opportunities to integrate and apply this knowledge as practitioners in a variety of field-based settings. Over time, prospective teachers are involved in a variety of field-based settings in which they interact with supervising or experienced teachers who function as mentors. During their preservice period, prospective teachers can explore, analyze, and question new ideas with support and encouragement.

The first few years of teaching present a very different period in the professional development process. Initial teaching assignments and support structures play a significant role in shaping beginning teachers' views of the profession as well as their commitment to it. The focus is on carrying out the responsibilities of teaching (e.g., planning for instruction, managing students' learning, responding to the changing needs of the learning environment) and involves a comprehensive application of what teachers have learned and experienced as part of their preservice programs. New issues are confronted, and knowledge and skills are built daily, more often within the context of the teaching environment than through formal continuing education.

As teachers of mathematics become more experienced, they move into another stage of their career; their collegial interactions increase and they assume new roles. Experienced teachers of mathematics become more involved in decisions about curriculum and staff-development programs. Indeed, the identification of staff-development needs and the development of programs to meet those needs can become part of the professional responsibilities of an experienced teacher. In addition, experienced teachers may become mentors to beginning or developing teachers. Experienced teachers' need for more formal continuing education ebbs and flows. As teachers reflect on their teaching and their students' understanding, they often become motivated to enhance their knowledge of mathematics, knowledge of students, and knowledge of teaching.

Assumptions

Several basic assumptions provide the foundation for the Standards detailed in this section:

1. *Principles and Standards for School Mathematics (Principles and Standards)* (NCTM 2000) provides the vision of mathematics education that is the basis for these education and professional development Standards. *Principles and Standards* is grounded in the belief that all students, prekindergarten through grade 12, should learn important mathematical concepts and processes with understanding. To participate in the realization of that goal, teachers need adequate professional preparation and continuing professional development geared toward developing their expertise in assessing and promoting students' understanding of mathematics. As a result, teacher development programs should focus on the knowledge, skills, understandings, and dispositions required to monitor, analyze, engage, and improve students' mathematical thinking.
2. Teachers' own learning experiences have a profound effect on their knowl-

edge of, beliefs about, and attitudes toward mathematics, students, and teaching. The many years that teachers spend as learners of pre-K–12 mathematics—consciously or unconsciously—provide them with images and models of what it means to teach and learn mathematics. They add to this background many other learning experiences, such as formal college preparation, clinical and field-based observation and practice teaching, the influence of school culture and colleagues within their immediate teaching environments, and in-service and advanced educational experiences. All these experiences convey messages about what constitutes appropriate methods of teaching and learning. Such powerful influences need to be addressed when helping teachers learn to teach in new ways.

3. Learning to teach is a process of integration. Although the Standards for the Education and Continued Professional Growth of Teachers of Mathematics separately address various components of teacher knowledge and practice, the final measure of success for any teacher is the integration of theory and practice. As they implement different instructional strategies, teachers should be discussing the research that supports their choices of those strategies. Instruction in mathematics content and pedagogy should be the result of coordinated efforts of mathematics education and mathematics faculty, so as to bridge what is often seen as a chasm between methods courses and content courses (CBMS 2001; Ball and Bass 2000a). Such integration is not easily achieved. Nonetheless, it is a goal to strive for in the improvement of teacher education.

Instruction in mathematics content and pedagogy should be the result of coordinated efforts of mathematics education and mathematics faculty, so as to bridge what is often seen as a chasm between methods courses and content courses.

4. The education of teachers of mathematics is an ongoing process. Teachers are in a constant state of "becoming." Being a teacher implies a dynamic and continuous process of growth that spans a career. Teachers' growth requires a commitment to professional development aimed at improving their teaching on the basis of increased experience, new knowledge, and awareness of educational reforms. This growth is deeply embedded in teachers' philosophies of learning, their attitudes and beliefs about learners and mathematics, and their willingness to make changes in how and what they teach. Teachers' growth potential can be enhanced or limited by the actions of others, including school administrators, educational policymakers, college and university faculty, parents, and the students themselves.

Being a teacher implies a dynamic and continuous process of growth that spans a career.

5. Mathematics teacher education must have grade-band-specific components. Although certain knowledge, skills, and abilities should be common to the preservice and continuing education of all teachers of mathematics, some elements of teacher development should be grade-band specific. For example, the breadth, depth, and scope of the mathematical content knowledge required for teaching differs for each grade band.

Organization

The statement of each of the five Standards in this section on education and professional growth is first elaborated with an explanation of its main ideas and

occasionally highlighted by quotations from mathematicians, mathematics educators, teachers, and students. Then, for each Standard, we follow with vignettes that show and extend these ideas through a variety of contexts related to the preservice and continuing education of teachers of mathematics. Drawn from transcripts, observations, and experiences, the vignettes are selected to illustrate a range of professional development opportunities and issues. The closing thoughts focus on issues pertinent to that Standard and, in some instances, include additional detail that elaborates as well as annotates.

Summary

The Standards in this section represent the National Council of Teachers of Mathematics' position on the essential components of the education and professional growth of teachers of mathematics. They comprise the threads that are woven into the fabric of successful mathematics teaching: personal experiences in contexts that model and value good mathematics teaching; ongoing development of knowledge about mathematics, students, and teaching; numerous and diverse opportunities to apply knowledge and experience through practice; and the gradual assumption of responsibilities for professional growth and change. Ideally, the fabric itself evolves and changes over time, reflecting the numerous stages in the career-long development of mathematics teachers.

Standard 1: Teachers' Mathematical Learning Experiences

Mathematics and mathematics education instructors in preservice and continuing education programs should model good mathematics teaching by—

- posing worthwhile mathematical tasks;
- creating supportive learning environments;
- expecting and encouraging intellectual risk-taking;
- engaging in collaborative mathematical discourse;
- enhancing discourse through the use of a variety of technologies, tools, and models;
- representing mathematics as an ongoing human activity; and
- affirming and supporting full participation and continued study of mathematics by all.

Elaboration

The experiences that mathematics teachers have while learning mathematics have a powerful effect on the education they provide their students. Prospective and practicing teachers spend many hours in mathematics and mathematics education classes, workshops, seminars, and other structured learning environments. Through those experiences, they develop ideas about what it means to teach mathematics, beliefs about successful and unsuccessful classroom practices, and strategies and techniques for teaching particular topics. The individuals from whom they are learning are role models who contribute to an evolving vision of what mathematics is and how mathematics is learned (Stigler and Hiebert 1999).

The experiences that mathematics teachers have while learning mathematics have a powerful effect on the education they provide their students.

The development of mathematics teacher candidates should include opportunities to examine the nature of mathematics, how mathematics should be taught, and how students learn mathematics. In particular, preservice teachers should observe and analyze a range of approaches to mathematics teaching and learning, focusing on the tasks, discourse, environment, and assessment that are associated with those approaches. Preservice teachers should work with a diverse range of students individually, in small groups, and in large-class settings to gain experience in facilitating learning experiences and assessing students' understanding in each of those settings. To help preservice teachers connect their learning experiences with their future roles as teachers, instructors of mathematics and mathematics education should explicitly address the "what, why, and how" of teaching—paying particular attention to tasks, environment, discourse, assessment—as detailed in the Standards for Teaching and Learning Mathematics.

This vision of teaching redirects mathematics instruction from a focus on presenting content through lecture and demonstration to a focus on active participation and involvement. An investigative nature is essential. Teachers cannot help students become seekers if they are not seekers themselves. Effective mathematics instructors for prospective teachers do not simply "deliver" content; rather, they facilitate learners' construction of their own knowledge of mathematics. Sometimes they stand back, allowing students to puzzle over, and come up with, their own solutions. Sometimes they push and lead, helping students reach particular reasoned conclusions. And sometimes they help students by modeling or telling. They do so through their choice of tasks and tools for instruction and their attention to the nature of the mathematical discourse that occurs in the learning environment they have created.

An investigative nature is essential. Teachers cannot help students become seekers if they are not seekers themselves.

Mathematics and mathematics education instruction should enable all learners to experience mathematics as a dynamic engagement in solving problems. These experiences should be designed deliberately to help teachers rethink their conceptions of what mathematics is, what kinds of activities take place in mathematics classes, and how mathematics is learned. Instruction should be organized around searching for solutions to problems and should include continuing opportunities to talk about mathe-matics. Working in groups is an excellent way for learners to explore ideas, develop mathematical arguments, conjecture, validate possible solutions, and identify

These experiences should be designed deliberately to help teachers rethink their conceptions of what mathematics is, what kinds of activities take place in mathematics classes, and how mathematics is learned.

connections among mathematical ideas. In such experiences, university instructors should guide explorations toward generalizations. Instructors should model the representation of mathematical ideas visually, in writing, and through dialogue and discussion and should help students develop facility with those representations as well.

Representations are crucial to the development of mathematical thinking, and through their use, mathematical ideas can be modeled, important relationships identified and clarified, and understandings fostered. Physical models, materials, calculators, and computers help provide the array of rich and substantive experiences needed to build deep and comprehensive knowledge of mathematical concepts and procedures. The experiences teachers have in such learning environments form expectations—implicitly or explicitly—of what constitutes good mathematics instruction. Such experiences provide the core from which teachers will eventually build learning environments for students.

Such instruction may require substantial changes in the philosophy and strategies of mathematics and mathematics education instructors at the university level who are involved in the preparation and continuing education of teachers of mathematics. Instructors need to experiment with new tasks, tools, and modes of classroom interaction and share and model new instructional strategies. Those changes necessitate collegial interaction and support, as well as participation in professional growth opportunities. Similarly, such changes necessitate changing the recognition and reward systems in colleges and universities. Also, school districts need to revise their perspectives on the design and structure of in-service support needed to effect substantive change. Finally, such changes place new expectations on teachers and students in their participation and engagement in learning. This climate challenges mathematics and mathematics education instructors to foster changes in their students' preconceived and generally traditional views about the way learning occurs.

Standard 1: Teachers' Mathematical Learning Experiences

Vignette

The following vignette demonstrates both the challenges and benefits of teaching preservice teachers in a student-centered environment. When preservice teachers first experience learning that requires them to make sense of mathematics rather than follow prescribed procedures, they may be surprised or even resentful that the rules of learning mathematics have suddenly changed. However, after adjusting to such an environment, preservice teachers often reap the benefits of deeper mathematical understandings and more powerful analytic capabilities.

1.1—*Changing Learning Expectations: An Instructor's Role*

Prospective middle or high school teachers are coming to the end of a yearlong mathematics course. By now they take it for granted that they are expected to make sense of mathematics, develop their own problems, make connections, and come up with further questions to extend their thinking.

This metamorphosis has not happened automatically. The mathematics professor has worked hard to achieve her goal of encouraging students to develop greater reliance on their abilities to make sense of mathematics. She has worked to help them become more active participants in, and creators of, their own learning. "It hasn't been easy to shift these students' expectations from wanting answers from the instructor to a point where they accept, and in fact demand, that they have a chance to make sense of a situation themselves."

When students confront their own perceptions about what it means to learn mathematics, their level of discomfort is an indication that change is happening. Early in the year, a preservice teacher who was interviewed about the class noted, "Up until the university placement exam, I just plugged numbers in and always got good grades. It had been a long time since I had math. I couldn't remember the way to do lots of the problems or appropriate formulas. I couldn't tackle problems if I didn't know the formula. I like to plug numbers into formulas. This math class is very upsetting. This is the first time I ever thought about why. I am going to have to learn to think about it [why] if I expect to teach math."

This math class is very upsetting. This is the first time I ever thought about why.

Now, later in the semester, preservice teachers are working on a research project that allows them to apply and extend the ideas they been have studying in probability and statistics. Because school students tend to focus on individual attributes rather than aggregate patterns in data (Konold and Higgins 2003), the instructor wants these preservice teachers to have lots of experience exploring real data, looking for patterns and relationships. The preservice teachers have been challenged to look at U.S. census data available on the Internet and make some observations about patterns in the data. They are to import the real data into a dynamic data-exploration software program, use the software to represent the data in ways that illustrate a relationship they have observed between two variables, and then present their findings to the class.

Preservice teachers find the dynamic data-exploration software easy to use, as it is designed for use with young children. Through this and other explorations in this class, the preservice teachers have also learned how much real data is readily available on the Internet.

The preservice teachers work independently and collaboratively to search for interesting relationships. Corey, a preservice teacher, says, "This is looking interesting! Would you look at that! People with higher levels of education seem to be more likely to relocate. Let's see if we can figure out the best way to show this relationship."

Corey's group members work together to explore different representations to determine which representations allow the apparent relationship to be seen at a glance. They explore box plots, scatterplots, and other representations and eventually settle on a ribbon chart for their presentation to the class.

Corey explains, "In our ribbon chart, the ribbon is divided along the horizontal axis into several categories of educational attainment. The width of each section of ribbon is determined by the number of cases in that category. The light-colored portion of ribbon represents people that relocated within each category, and the people who didn't relocate are represented by the darker-colored portion of ribbon. So you can see that as the educational attainment increases, so does the relative number of relocators."

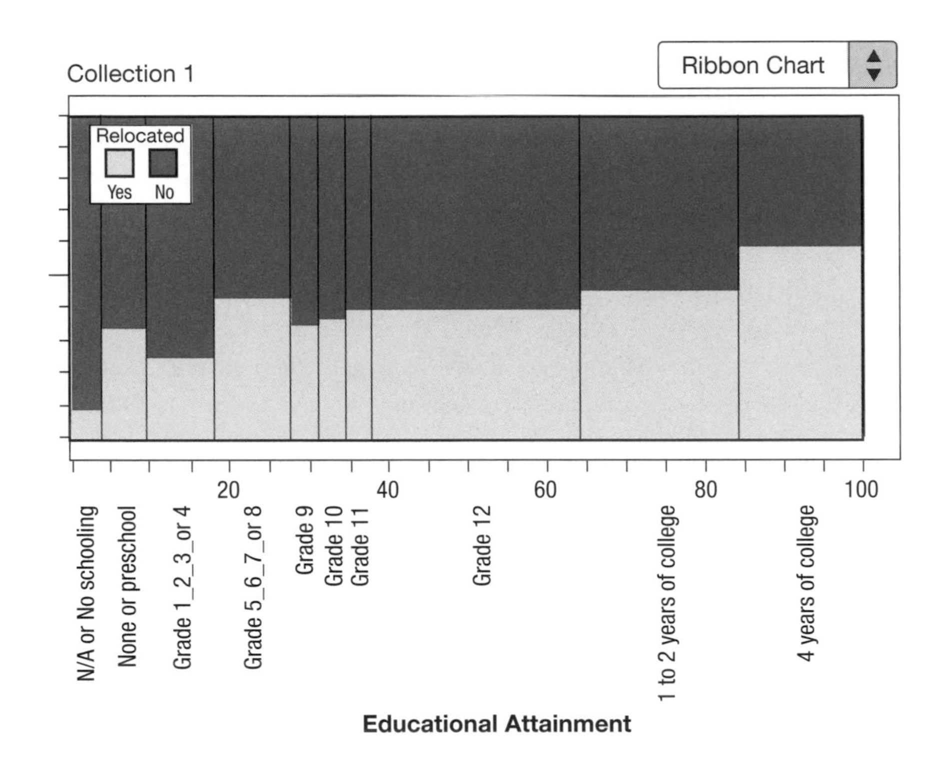

Jennie, one of Corey's group mates, notices, "Some of the light-colored bars aren't much higher on the right side of the chart than those in the center of the chart. I'm not so sure that you can necessarily predict whether a person will relocate based on their education level." Claire, another student, asks, "What do we know about the ages of the people in the data?" Charlotte adds, "That's a good question, Claire. Maybe people are more likely to relocate as they get older. Little kids would probably not be very likely to relocate." The instructor is delighted by the students' comments. Jennie has set the stage for introducing ways to measure the strength of the conjectured relationship, and Claire and Charlotte have identified a potential lurking variable. The students' observations will provide a nice lead-in to a discussion about the need to develop statistical tools for dealing with multivariable data sets.

Students find this kind of involvement in the class different from other experiences they have had in mathematics. Max, a student in the class, pointed out a contrast between problems in this class and assignments in a typical math class:

I used to think that math was just busy work. Do thirty problems on finding the mean, median, and mode of some fake data sets. It becomes mindless by the thirtieth problem, and you don't really know why you're performing these calculations. But this class doesn't have any busywork. We use a lot of real data so I know why what we're learning is important. I can also see how what I'm learning in this class is a model for how I can keep the interest and attention of the middle or high school students I'll be teaching.

I can also see how what I'm learning in this class is a model for how I can keep the interest and attention of the middle or high school students I'll be teaching.

Vignette

In the next vignette, a university mathematician incorporates technology into a calculus course to help students discover the relationship between a graph and its derivative. The use of technology enabled students to share some of their insights and misconceptions about relationships between variables and gave them a powerful model to draw on throughout the course. The vignette as a whole is another example of how instructors can model mathematics teaching strategies that are also appropriate to use with school students.

1.2—*Modeling Instructional Strategies: Using Technology in Calculus*

Hallie Foster, professor of mathematics at a midsized university, integrates technological tools with numerical, graphical, and symbolic manipulation capabilities into a calculus course that emphasizes concepts, principles, and applications. Hallie uses the technology to engage students in active exploration and development of concepts.

To help students develop a sense of the relationship between a sine function and its derivative, Hallie devises a classroom demonstration in which data on the position of a swinging pendulum held directly over a Calculator Based Ranger (CBR) or motion detector is collected for several seconds and stored in a graphing calculator.

First, Hallie demonstrates the motion of the swinging pendulum and asks her students to predict what a graph of position versus time would look like for several complete revolutions, ignoring, for now, any effects of gravity. After viewing some of the students' predictions, Hallie asks several students to share their predictions and reasoning with the class. Roger sketches a figure that goes up and down in straight-line segments.

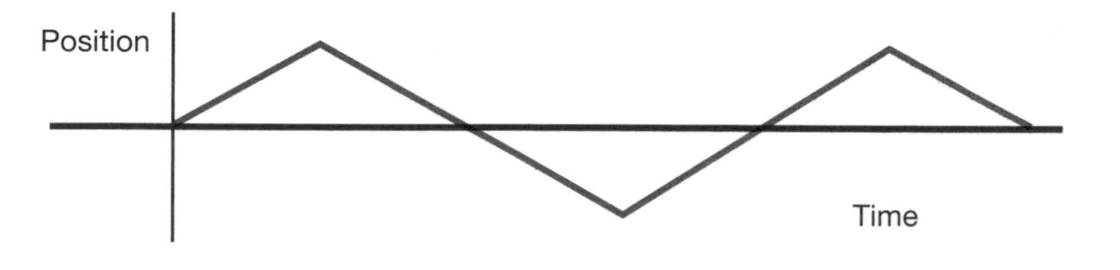

He says, "Well, the graph goes up as the pendulum moves away from the motion detector, then comes down as the pendulum comes back to the motion detector. It continues coming down as the pendulum passes the motion detector until the pendulum reaches its farthest point. Then the graph goes back up again as the pendulum heads back in the other direction."

Sharon shares her graph, which has the same general shape, although the up and down portions are curved rather than straight line segments.

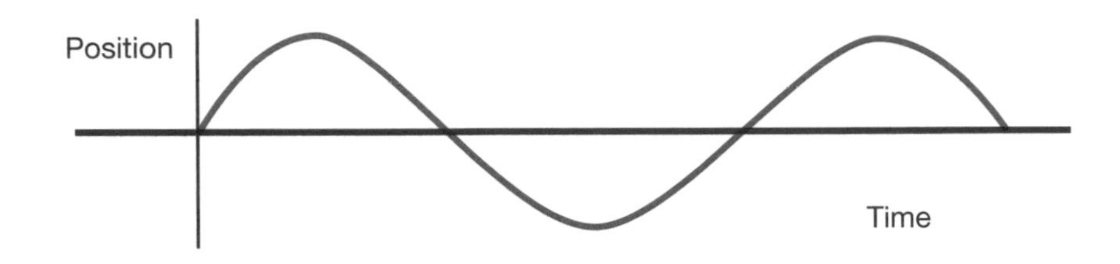

Sharon says, "My graph is a lot like Roger's except it's curved. I think that the pendulum's position changes faster when its closer to the bottom (motion detector) and more slowly when it's about to switch direction. So the graph has to be steeper at the beginning of the cycle and less steep when it's farthest away from the motion detector."

Hallie then shares a graph of position versus time generated by the data stored in the graphing calculator. She asks students to reflect on the explanations they have heard and the data they have seen represented in a graph and think about what makes sense in terms of the physical situation they have seen.

Next Hallie asks students to sketch a graph of velocity versus time for the swinging pendulum. Again she has students share their predictions, and she uses the graphing calculator to show the velocity graph for the stored data. Several "oohs" and "ahhs" are heard from the students. Bev says, "It looks like the same graph only shifted over." Hallie asks students to discuss in their groups whether it is the same graph and to make some conjectures about what types of equations might be used to fit these data.

Later in the semester, when Hallie wants to have students recall the relationship between position and velocity functions, she reminds students of the pendulum experiment. As a result of the pendulum and other technology-based experiences in Dr. Foster's class, the preservice teachers in her class have some good examples of how technology can be used as a device to enhance learning rather than merely as a substitute for paper-and-pencil computations.

Closing Thoughts
Standard 1: Teachers' Mathematical Learning Experiences

This standard addresses the role of formal mathematics instruction in mathematics teacher development. Through the experiences of learning mathematics, prospective and practicing teachers develop many of their core beliefs about how mathematics is learned and, therefore, how it should be taught. If teachers are to learn how to create a positive environment that promotes collaborative problem solving, incorporates technology in a meaningful way, invites intellectual exploration, and supports student thinking, they themselves must experience learning in such an environment.

If teachers are to learn how to create a positive environment that promotes collaborative problem solving, incorporates technology in a meaningful way, invites intellectual exploration, and supports student thinking, they themselves must experience learning in such an environment.

Standard 2: Knowledge of Mathematical Content

Teacher development programs should ensure that mathematics teachers are fluent in the language of mathematics and have broad and deep knowledge of mathematical content, processes, and contexts, including—

- mathematical concepts and procedures and the connections among them;
- multiple representations of mathematical concepts and procedures;
- ways to reason mathematically, solve problems, and communicate mathematics effectively at different levels of formality;
- the cultural contexts for mathematics, including the contributions of different cultures toward the development of mathematics and the role of mathematics in culture and society;
- the evolving nature of mathematical practice and instruction resulting from the availability of technology; and
- the relationship of school mathematics to the discipline of mathematics, to other fields of study, and to mathematical applications.

Elaboration

Knowledge of both the content and processes of mathematics is an essential component of teachers' preparation for the profession. Teachers' comfort with, and confidence in, their own knowledge of mathematics affects both what they teach and how they teach it. Their conceptions of mathematics shape their choices of worthwhile mathematical tasks, the kinds of learning environments they create, and the nature of the discourse in their classrooms.

Teachers' comfort with, and confidence in, their own knowledge of mathematics affects both what they teach and how they teach it.

Knowing mathematics includes understanding specific concepts and procedures; being fluent with its terminology, syntax, and notation; as well as having mathematical "habits of mind," such as experimenting, visualizing, conjecturing, and abstracting (Cuoco,

Goldenberg, and Mark 1996; Levasseur and Cuoco 2003). Mathematics involves the study of concepts and properties of numbers, geometric objects, functions, and their uses—identifying, counting, measuring, comparing, locating, describing, constructing, transforming, and modeling. The relationships and recurring patterns among those objects and the operations on those objects lead to the building of such mathematical structures as number systems, groups, or vector spaces and the study of the similarities and differences among those structures. Mathematical concepts and structural properties are used to create powerful algorithms or procedures for solving whole classes of problems.

Such knowledge should not be developed in isolation. The abilities to identify, define, and discuss concepts and procedures; to develop an understanding of the connections among them; and to appreciate the relationship of mathematics to other fields are crucially important. Mathematics both arises out of and influences continued development of other fields. Advances in mathematical thought spur advances in physics. Advances in computer science raise new mathematical problems to be solved. By interacting with mathematicians, preservice or in-service teachers can begin to understand the culture of mathematical practice and to appreciate mathematics as the complex and ever-changing field that it is.

Knowing mathematics also involves the larger context of mathematical discourse in which specific concepts and procedures are embedded. Discourse in mathematics centers on examining patterns, abstracting, generalizing, and making convincing mathematical arguments. It involves the role of definitions, examples, and counterexamples and the use of assumptions, evidence, and proof. Framing mathematical questions and conjectures, constructing and evaluating arguments, making connections, and communicating mathematical ideas all are important aspects of mathematical discourse. Engaging in mathematical discourse is central to how teachers come to know mathematics; to develop confidence in their own abilities to do mathematics; and to become aware of, and have an appreciation for, the place of discourse in the discipline of mathematics. Learning takes place with exchanges of information and ideas. We do not learn in isolation. As much learning occurs in small groups as takes place at an individual desk.

Teachers need opportunities to construct mathematics for themselves and not merely accept the results of others' constructions.

As part of the environment of discourse, the development of abilities in mathematical reasoning and problem solving is essential. Mathematical reasoning involves an interplay of intuitive, informal exploration and formal, systematic proof. All too often, the formal written record of mathematics is all that teachers study. The struggles, the false starts, and the informal investigations that lead to an elegant proof frequently are missing. Teachers need opportunities to construct mathematics for themselves and not merely accept the results of others' constructions. In addition, teachers need to interact with others in posing and solving problems to develop a repertoire of problem-solving strategies.

As an ongoing product of human activity, mathematics is a dynamic and pending system of connected principles and ideas constructed through exploration and investigation. Developing such a perspective includes an appreciation for the historical and cultural contributions made to the development of mathematics. It provides a provocative backdrop that may be useful in motivating students as they approach new subject matter and in encouraging the full participation and continued study of mathematics by all students.

Mathematics is a dynamic discipline that continues to grow and expand in its uses in our culture. Teachers will be called on to adapt to curriculum changes that this growth will entail. The study of contributions made to the development of mathematics by different cultures should provide teachers with resources to use in motivating students as they approach new subject matter.

Technology is a vital force in learning, teaching, and doing mathematics, providing new approaches for solving problems and influencing the kinds of questions that are investigated. It should play a significant role in the teaching and learning of mathematics. Technology can be used in a variety of ways to enhance and extend mathematics learning and teaching. By far the most promising are in the areas of problem posing and problem solving in activities that permit students to design their own explorations and create their own mathematics.

Technology changes the nature and emphasis of the content of mathematics as well as the pedagogical strategies used to teach mathematics. Indeed, one central issue revolves around the fact that some of the computational procedures that have formed the basis for mathematics courses at all levels are no longer essential. When performing computational and representational procedures by hand, students often lose sight of mathematical insights or discoveries as they become mired in the mechanics of producing the results. The introduction of technology necessitates reassessing the emphasis on algorithmic skills, algebraic as well as arithmetic, and paying greater attention to the power of dynamic graphic representation. Indeed, appropriate use of technology—computers and calculators—gives students access to powerful new ways to explore concepts at a depth that has not been possible in the past.

Technology changes the nature and emphasis of the content of mathematics as well as the pedagogical strategies used to teach mathematics.

Central to the preparation for teaching mathematics is the development of a deep understanding of the mathematics of the school curriculum and how it fits within the discipline of mathematics (Ma 1999; American Mathematical Society Resource Group on NCTM Standards 1998; Conference Board of the Mathematical Sciences [CBMS] 2001). Too often, teachers' knowledge of the content of school mathematics is assumed to be in place by the time they complete their own K–12 learning experiences. Yet investigation into how teachers hold their knowledge of mathematics has shown that they often do not grasp the underlying concepts of the school mathematics they teach (Ma 1999; Mathematical Sciences Education Board [MSEB] 2001). Research has also shown that simply taking further standard college mathematics courses does not adequately address their need (National Research Council 2001; CBMS 2001). Teachers need opportunities in university mathematics classes to revisit school mathematics topics in ways that will allow them to develop deeper understandings of the subtle ideas and relationships that are involved between and among concepts.

Too often, teachers' knowledge of the content of school mathematics is assumed to be in place by the time they complete their own K–12 learning experiences.

Such opportunities should include developing broad understandings of significant mathematics concepts and how they are related to other parts of the curriculum. At all levels, teachers need to see the "big picture" of mathematics across the elementary, middle, and high school years. To use a geographic analogy, teachers need to have a mental roadmap that shows the major cities (curriculum topics) and the roads (mathematical connections) among them. Such a mathematical map should also highlight the importance of connections between mathematics and other school subjects and between mathematics and situations in nonschool settings out of which mathematics arises or in which it is applied.

Common Experiences in the Mathematical Education of Teachers

Common experiences should be ingredients in the ways teachers of mathematics build and extend their knowledge of mathematics. Regardless of the context, the following processes, as suggested in *Principles and Standards for School Mathematics* (NCTM 2000), should be prominent in those experiences:

- Problem solving
- Reasoning and proof
- Communication
- Connections (both within mathematics and across disciplines)
- Representations

In addition, mathematical experiences for all teachers should incorporate—

- equity, in the form of high expectations and strong support for all learners;
- curriculum that is coherent and comprehensive;
- teaching that is competent and caring;
- learning that is focused on understanding and using mathematics;
- assessment that monitors, enhances, and evaluates learning to guide instruction; and
- technology use that enhances mathematical understanding.

These experiences can occur in mathematics courses, workshops, conferences, or other professional growth activities. In the process of constructing and developing such experiences, appropriate attention to, and use of, mathematical modeling and technology should be included to enhance the teaching and learning of the mathematical ideas. To that end, teachers should become familiar with powerful numerical, symbolic, and graphical tools, including manipulatives and electronic technologies, that provide for the exploration, investigation, and application of mathematics. Those tools should be incorporated in instruction and used for assignments whenever such inclusion can add to students' insight and understanding.

Because teachers are expected to incorporate the mathematical processes of problem solving, reasoning, communicating, connecting, and representing into their teaching, making those processes part of the teachers' mathematical experiences is imperative. The next section makes more explicit the vision of how mathematical processes must play a role in the preparation and continuing education of teachers.

Because teachers are expected to incorporate the mathematical processes of problem solving, reasoning, communicating, connecting, and representing into their teaching, making those processes part of the teachers' mathematical experiences is imperative.

Process Knowledge for Teachers of Pre-K–12 Mathematics

Knowledge of Mathematical Problem Solving

Teachers and teacher candidates must know, understand, and apply the process of mathematical problem solving. As they study mathematics themselves, they must learn to apply and adapt a variety of appropriate strategies to solve problems that arise in mathematics and in other contexts. They must build new mathematical knowledge through problem solving and monitor as well as reflect on the process of mathematical problem solving.

Knowledge of Reasoning and Proof

Teachers and teacher candidates need to reason about, construct, and evaluate mathematical arguments and develop an appreciation for mathematical rigor and inquiry. They should learn to recognize reasoning and proof as fundamental aspects of mathematics, make and investigate mathematical conjectures, develop and evaluate mathematical arguments and proofs, and select and use various types of reasoning and methods of proof.

Knowledge of Mathematical Communication

Teachers and teacher candidates must learn to communicate their mathematical thinking coherently and clearly, both orally and in writing. They must learn to use the language of mathematics to express ideas precisely, to organize mathematical thinking through communication and analysis, and to evaluate the mathematical thinking and strategies of others.

They must learn to use the language of mathematics to express ideas precisely, to organize mathematical thinking through communication and analysis, and to evaluate the mathematical thinking and strategies of others.

Knowledge of Mathematical Connections

Teachers and teacher candidates must learn to recognize, use, and make connections between and among mathematical ideas and in contexts outside mathematics to build mathematical understanding. They must be able to demonstrate how mathematical ideas interconnect and build on one another to produce a coherent whole.

Knowledge of Mathematical Representation

Teachers and teacher candidates need to be exposed to varied representations of mathematical ideas that support and deepen mathematical understanding. They need to use representations to model and interpret physical, social, and mathematical phenomena. They need to create and use representations to organize, record, and communicate mathematical ideas as well as be able to select, apply, and translate among mathematical representations to solve problems.

They must have a sense of how the mathematics they teach fits into the larger body of school mathematics, and their mathematical knowledge must go well beyond the scope of what they expect to teach.

In addition to the knowledge and experiences involving the processes cited above, teachers of, and teacher candidates for, pre-K–12 mathematics must understand the role and value of each of the NCTM Principles: Equity, Curriculum, Teaching, Learning, Assessment, and Technology. Those Principles, interwoven throughout this *Standards* document, form the basis for high-quality mathematics instruction for students and teachers. In addition, teachers and teacher candidates must have a deep, connected understanding of the mathematics they teach. They must have a sense of how the mathematics they teach fits into the larger body of school mathematics, and their mathematical knowledge must go well beyond the scope of what they expect to teach (CBMS 2001; NCTM 2006). The next section outlines the content knowledge required by grades Pre-K–12 teachers in the spirit of these recommendations.

Content Knowledge

The study of mathematics must be viewed as a collection of related concepts, procedures, and intellectual processes. This view requires a multicourse approach in which topics are integrated with one another in rich, problem-based contexts. In other words, mathematics must be treated as a unified whole. In such an environment, preservice teachers would have an opportunity to develop the sense that the whole (of mathematics) is greater than the sum of the parts. That perspective should be reflected consistently in content courses, methods courses, and field experiences.

The design and implementation of mathematics courses for teachers must be consistent with the outcomes that are expected of teachers in school-level education.

The design and implementation of mathematics courses for teachers must be consistent with the outcomes that are expected of teachers in school-level education. In particular, mathematics instructors should promote equity in their treatment of content and in their dealings with learners. They should teach mathematical content in a way that embeds it in its cultural and historical contexts so as to highlight contributions by individuals from diverse cultural backgrounds and experiences and to emphasize the role of mathematics in society. Instructors should use worthwhile mathematical tasks to engage learners in mathematical processes and to stimulate interest in learning mathematics. They should explore those tasks with their students in a classroom environment that invites participation, reflects collaborative negotiation of meaning, and is conducive to intellectual risk taking.

Classroom discourse should involve both instructors and learners and should be used as a vehicle for making sense of mathematics. Likewise, technological tools, such

as spreadsheets, interactive geometry software, computer algebra systems, dynamic statistical packages, graphing calculators, data-collection devices, and presentation software, should be integrated into learning experiences to enhance mathematical understanding, not be a substitute for it. The extent to which teachers and teacher candidates are learning sound and significant mathematics with understanding should be determined by analysis of the results of a variety of formal and informal assessments. The quality of the tasks, discourse, and environment should also be assessed to guide and improve instruction.

The discussion that follows identifies the mathematics content needed by all teachers in grades pre-K–12 and based predominantly, but not exclusively, on the National Council for Accreditation of Teacher Education (NCATE) content requirements as stipulated by the NCTM Specialty Program Area (http://www.nctm.org/about/ncate/). The NCATE recommendations are also consistent with many recommendations made by the Teacher Education Accreditation Council (TEAC), the National Board on Professional Teaching Standards, and others that license or certify teachers or that evaluate teaching or teacher education programs. The discussion first presents the mathematics needed by all mathematics teachers, followed by additional mathematics needed by middle school (grades 6–8) mathematics teachers, and finally, still additional mathematics that is needed by those who plan to teach mathematics in secondary school (grades 9–12). These requirements ensure that teachers at all grade levels have not only a thorough understanding of the mathematics they are teaching but also a perspective on how that mathematics fits into the scheme of all school-level mathematics.

The diagram below illustrates the embedded and connected nature of content knowledge needed by mathematics teachers at various levels. At the core, the inner circle represents knowledge required by all teachers: early childhood (Pre-K–2), elementary school (3–5), middle grades (6–8), and secondary school (9–12). The ring surrounding the core represents additional content knowledge required by middle level (6–8) and secondary school (9–12) teachers. Finally, the outer ring represents additional content knowledge required by secondary school (9–12) teachers.

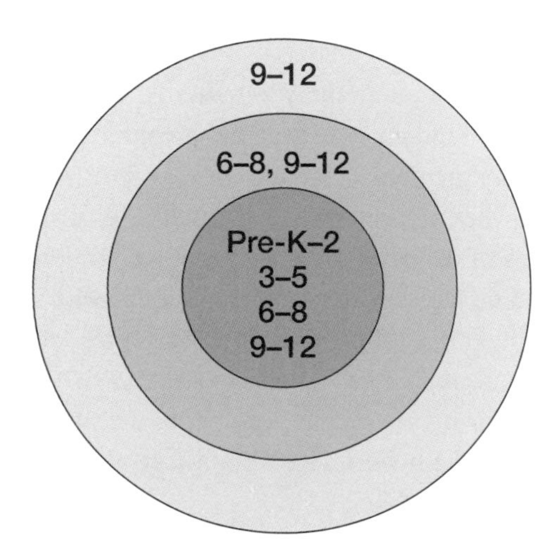

Content Knowledge Needed by All Mathematics Teachers: Early Childhood (Pre-K–2), Elementary School (3–5), Middle Grades (6–8), and Secondary School (9–12)

Knowledge of Number and Operation

All mathematics teachers and teacher candidates must be able to demonstrate mathematical proficiency with respect to number and operations, including a conceptual understanding of numbers, ways of representing numbers, relationships among numbers and number systems, and the meanings of operations.

They should be able to develop the meaning and interrelations of addition, subtraction, multiplication, and division and provide multiple models for, and applications of, whole-number operations. They should recognize the meaning and use of place value in representing whole numbers and finite decimals, comparing and ordering numbers, and understanding the relative magnitude of numbers. They should demonstrate proficiency in, and understanding of, multidigit computation using a variety of standard and nonstandard algorithms, mental mathematics, and computational estimation. They should be able to analyze integers and rational numbers, their relative size, and how operations with whole numbers extend to integers, rational numbers, and real numbers. They should be able to form convincing arguments about number patterns and relationships and demonstrate knowledge of the historical development of number and number systems, including contributions from diverse cultures.

Knowledge of Algebra

All mathematics teachers and teacher candidates must be able to emphasize relationships among quantities, including functions, ways of representing mathematical relationships, and the analysis of change.

They should be able to explore and analyze patterns, relations, and functions. They should be able to recognize and analyze mathematical structures, investigate equality and equations, and use mathematical models to represent quantitative relationships. They should be able to identify and use commutativity, associativity, distributivity, identities, and properties of inverses and connect those properties with computational algorithms. They should be facile with graphical, numerical, and symbolic representations and understand the connections among those representations. They should be able to analyze and represent change in various contexts, such as in rates and ratios. They should be able to form convincing arguments within an algebraic system and demonstrate knowledge of the historical development of algebra, including contributions from diverse cultures.

Knowledge of Geometry

All mathematics teachers and teacher candidates must be able to use spatial visualization and geometric modeling to explore and analyze geometric shapes, structures, and their properties.

They must be able to use visualization, the properties of two- and three-dimensional shapes, and geometric modeling. They should be able to build and manipulate representations of two- and three-dimensional objects using concrete models, representational paper-and-pencil drawings, and interactive geometry software. They should be able to specify locations and describe spatial relationships using synthetic, coordinate, and transformational geometry. They should understand the relationships among the concepts of symmetry, congruence, and similarity. They should have a broad technical vocabulary and understand the value and role of mathematical definitions. They should be able to form convincing arguments within a geometric system and demonstrate knowledge of the historical development of plane and spherical geometries, including contributions from diverse cultures.

Knowledge of Data Analysis and Probability

All mathematics teachers and teacher candidates should be able to demonstrate an understanding of concepts and practices related to data analysis and probability.

They should be able to design investigations that can be addressed by creating data sets and collecting, organizing, and displaying relevant data. They should use appropriate parametric and nonparametric statistical methods and technological tools to analyze data and describe shape, spread, and center. They should understand issues of variability, sampling, and inference. They should be able to choose among various representations and summary statistics to communicate conclusions. They should be able to make judgments under conditions of uncertainty and be familiar with concepts of likelihood and randomness. They should explore the concepts of empirical and theoretical probabilities using technology and manipulative-based simulations as well as analysis of the underlying sample space. They should be able to use statistical or probabilistic reasoning to form convincing arguments and demonstrate knowledge of the historical development of probability and statistics, including contributions from diverse cultures.

Knowledge of Measurement

All mathematics teachers and teacher candidates should be able to apply and use measurement concepts and tools.

They should be able to select and use appropriate measurement units, techniques, and tools, and be able to recognize and apply measurable attributes of objects and the units, systems, and processes of measurement. They should be familiar with both standard

(English and metric) and nonstandard units of measurement. They should be able to compute, apply, and connect the various measurement concepts, such as length, area, volume, perimeter, surface area, time, temperature, angle, weight, and mass. They should understand that all measurements are necessarily approximations and that the choice of units affects precision. They should be able to derive measurement formulas and construct convincing arguments on the basis of measurement concepts. As a way of understanding measurement units and processes, they should demonstrate knowledge of the historical development of measurement and measurement systems, including contributions from diverse cultures.

Additional Content Knowledge Needed by Mathematics Teachers in the Middle Grades (6–8) and Secondary School (9–12)

In addition to the content knowledge required of all teachers of mathematics described above, middle grades and secondary school teachers also need the following knowledge, skills, and understandings.

Knowledge of Number and Operation

Mathematics teachers and teacher candidates specializing in middle-grades or secondary school instruction should be able to demonstrate mathematical proficiency with respect to number and operations, including a conceptual understanding of numbers, ways of representing numbers, relationships among numbers and number systems, and meanings of operations.

They should be able to represent and work with other number systems, such as complex numbers, clock arithmetic, modular systems, and matrices. They should be able to make sense of large and small numbers and use scientific notation to represent and compute with such numbers. They should have a deep understanding of proportional reasoning and of direct and inverse relations, and be able to represent those relationships in tabular, graphical, and symbolic ways. They should be able to provide equivalent representations of fractions, decimals, ratios, and percents. They should be able to model operations using rational numbers and connect those models with operations involving whole numbers. Likewise, they should be able to explain the distinctions among whole numbers, integers, rational numbers, and real numbers and be able to determine whether the field axioms hold for each set of numbers and operations. They should apply the fundamental ideas of number theory, including concepts associated with prime and composite numbers. They should be able to construct convincing arguments within the domain of basic number theory and understand the cultural context and historical development of additional concepts included at that level.

Knowledge of Algebra

Mathematics teachers and teacher candidates specializing in middle-grades or secondary school instruction should be able to identify and work with families of functions, understand the connections between arithmetic and algebra, manipulate algebraic representations with understanding, and use algebraic structures to model physical phenomena.

They should understand the role of variables, expressions, equations, and inequalities in representing quantities and relationships. They should be familiar with the attributes and patterns of change associated with families of functions, such as linear, quadratic, and exponential functions. They should know how to represent proportional relationships using linear functions. They should recognize algebra as a symbolic language and use the tools of algebra as a means of representing generalized arithmetic relationships. They should be able to use algebraic representations to develop models of physical phenomena and as a tool for solving problems. They should be able to construct and interpret representations of algebraic relationships using such instructional tools as manipulatives and dynamic algebra software. They should be able to use algebraic reasoning to provide justifications for algorithms or algebraic equivalences.

They should be able to use algebraic reasoning to provide justifications for algorithms or algebraic equivalences.

Knowledge of Geometry

Mathematics teachers and teacher candidates specializing in middle-grades or secondary school instruction should be able to use spatial visualization and geometric modeling to explore and analyze geometric shapes, structures, and their properties.

They should be able to demonstrate knowledge of core concepts and principles of Euclidean geometry in two and three dimensions. They should be able to investigate patterns and properties of geometric shapes. They should be able to formulate conjectures on the basis of their observations, then prove or disprove their conjectures. They should be able to perceive and represent two- and three-dimensional objects from different perspectives. They should be familiar with the proportional relationships associated with similar figures. They should know which properties of a figure are preserved under various transformations. They should be able to demonstrate how similar or congruent figures are generated as a result of rotations, reflections, translations, or dilations. They should be aware of connections among geometry, art, nature, and culture.

Knowledge of Data Analysis and Probability

Mathematics teachers and teacher candidates specializing in middle-grades or secondary school instruction should be able to demonstrate an understanding of concepts and practices related to experimental design, data analysis, and probability.

They should use appropriate statistical methods and technological tools (especially software designed for use by middle-grades and high school students) to analyze data and describe shape, spread, and center, as well as investigate, interpret, and construct representations for conditional and geometric probability and for bivariate data.

They should be able to design investigations and to collect data through random sampling or random assignment to treatments to answer specific questions. They should understand issues of bias and look for ways to reduce bias in experimental design. They should be able to use a variety of ways to display the data and interpret data representations. They should be able to recognize patterns in data displays as well as departures from patterns, such as extreme values and outliers. They should identify misuses of statistics and invalid probabilistic claims. They should use appropriate statistical methods and technological tools (especially software designed for use by middle-grades and high school students) to analyze data and describe shape, spread, and center, as well as investigate, interpret, and construct representations for conditional and geometric probability and for bivariate data. They should be able to make inferences and decisions based on statistical and probabilistic reasoning. They should be aware of the uses of statistics and probability in a variety of applied fields.

Knowledge of Measurement

Mathematics teachers and teacher candidates specializing in middle-grades or secondary school instruction should be able to select, apply, and analyze measurement concepts and tools.

They should be able to make appropriate choices of handheld or electronic measurement tools and use tools carefully and with a reasonable degree of accuracy. They should have a sense of the degree of error associated with various tools and the potential effect of measurement errors on formulas or calculations to which they are applied. They should be able to employ estimation as a way of predicting reasonable measurements including the outcomes of conversions between measurements. Likewise, they should be able to use dimensional analysis as a way of evaluating the reasonableness of a problem solution. They should understand the relationships between measurements of various dimensions and volumes and understand how change in a measurement of one dimension affects the values of measurements in other dimensions.

Knowledge of Calculus

Mathematics teachers and teacher candidates specializing in middle-grades or secondary school instruction should be able to demonstrate a conceptual understanding of limit, continuity, differentiation, and integration and a thorough background in the techniques and application of calculus.

They should be able to describe and interpret symbolic, graphical, and tabular representations of limits. They should understand the definitions of derivatives and integrals as particular instances of limits and be able to represent those definitions graphically. They should be able to identify both continuous and discontinuous functions. They should be able to identify graphical function attributes associated with the derivative. They should be able to explain and apply the concept of optimization. They should know how calculus is applied in medicine, sciences, manufacturing, business, and other fields. They should demonstrate knowledge of the historical development of calculus, including contributions from diverse cultures.

Knowledge of Discrete Mathematics

Mathematics teachers and teacher candidates specializing in middle-grades or secondary school instruction should be able to apply the fundamental ideas of discrete mathematics in the formulation and solution of problems.

They should be able to demonstrate a conceptual understanding of the fundamental ideas of discrete mathematics, such as finite graphs, trees, and combinatorics. They should know how to use technological tools to apply the fundamental concepts of discrete mathematics. They should demonstrate knowledge of the historical development of discrete mathematics, including contributions from diverse cultures.

They should demonstrate knowledge of the historical development of discrete mathematics, including contributions from diverse cultures

Additional Content Knowledge Needed by Secondary School Teachers (9–12)

In addition to the content knowledge required of all teachers of mathematics described in the foregoing and the additional knowledge required by both middle-grades and secondary school teachers, secondary school teachers also need the following knowledge, skills, and understandings.

Knowledge of Number and Operations

Mathematics teachers and teacher candidates specializing in secondary school instruction should be able to demonstrate mathematical proficiency with respect to number and operations, including a conceptual understanding of numbers, ways of representing numbers, relationships among numbers and number systems, and meanings of operations.

They must be able to analyze and explain the mathematics that underlies the procedures used for operations involving integral, rational, real, and complex numbers. They should be able to compare and contrast properties of numbers and number systems. In particular, they should recognize matrices and vectors as systems that have some of the properties of the real number system. They should have a good grasp of number theory and its connections with school mathematics. They should also be familiar with applications of number theory, such as coding and computing.

Knowledge of Algebra

Mathematics teachers and teacher candidates specializing in secondary school instruction should be able to demonstrate a mathematical proficiency with relationships among quantities, including functions, ways of representing mathematical relationships, and the analysis of change.

They should be able to analyze patterns, relations, and functions of one and two variables. They should have extensive understanding of linear algebra, including vector spaces and linear transformations. They should know how vector space ideas connect with high school mathematics, and they should explore connections between linear algebra and analytic geometry. They should know how to apply the major concepts of abstract algebra to justify algebraic operations and formally analyze algebraic structures. They should have experience working with rings, integral domains, and fields and should explore the connections with high school mathematics content. They should investigate topics that help build connections among various content domains, such as isometry groups of regular polygons. They should be able to demonstrate knowledge of the historical development of linear algebra and abstract algebra, including contributions from diverse cultures.

Knowledge of Geometry

Mathematics teachers and teacher candidates specializing in secondary school instruction should be able to use spatial visualization and geometric reasoning to understand the structure, patterns, and applications of geometries.

They should be able to demonstrate knowledge of core concepts and principles of Euclidean and non-Euclidean geometries from both formal and informal perspectives. They should be able to exhibit knowledge of the role of axiomatic systems and proofs in the content of each geometric system. They should be able to specify locations and describe spatial relationships using coordinate geometry, vectors, and other representational systems. They should be familiar with connections between algebra and geometry, especially as it is embedded in the high school curriculum. They should be familiar with a variety of geometric applications, such as tiling, robotics, and computer graphics. They should be able to demonstrate knowledge of the historical development of geometries and various applications, including contributions from diverse cultures.

Knowledge of Data Analysis and Probability

Mathematics teachers and teacher candidates specializing in secondary school instruction should be able to demonstrate an understanding of concepts and practices related to data analysis, statistics, and probability.

They should know how to estimate population means and proportions. They should know how to apply a chi-squared test for goodness of fit. They should be able to perform regression analysis and understand the geometric and analytic derivation of this procedure. They should know how to use and interpret analysis of variance. They should be able to deal with a large data set and to apply and interpret statistical tests that make sense for the data they are dealing with. They should understand basic probabilistic concepts, such as independence, conditional probability, Bayes' theorem, common discrete and continuous probability distributions, and the central limit theorem. They should understand the probabilistic underpinnings of statistical inference, such as common sampling distributions, point estimation, hypothesis testing, and confidence intervals. In addition, they should be able to demonstrate knowledge of the historical development of these statistics and probability topics, including contributions from diverse cultures.

Knowledge of Measurement

Mathematics teachers and teacher candidates specializing in secondary school instruction should be able to apply and use measurement concepts and tools.

Knowing measurement includes understanding specific concepts and procedures as well as the process of doing mathematics. Teachers should embrace technology as an essential tool for teaching and learning measurement. They should recognize the common representations and uses of measurement. They should apply appropriate techniques, tools, and formulas to determine measurements and their application in a variety of contexts, such as the determination of the volume of non-uniform solids or solids formed by rotations. They should complete error analysis through determining the reliability of the numbers obtained from measures. They should be able to demonstrate knowledge of the historical development of measurement and measurement systems, including contributions from diverse cultures.

Knowledge of Calculus

Mathematics teachers and teacher candidates specializing in secondary school instruction should understand concepts of univariate and multivariate calculus and be familiar with applications in these contexts.

They should be able to demonstrate a conceptual understanding of, and procedural facility with, basic calculus concepts; apply concepts of function, geometry, and trigonometry in solving problems involving calculus; use the concepts of calculus and mathematical modeling to represent and solve problems taken from real-world contexts; and use technological tools, including computer algebra systems and dynamic software environments, to explore and represent fundamental concepts of calculus. They should understand the connections between univariate and multivariate calculus. They should know how to solve simple differential equations and recognize problem contexts that reflect differential relationships. They should know how to compute with sequences and series and be familiar with problems in the secondary school curriculum that can be modeled using those constructs. In addition, they should be able to demonstrate knowledge of the historical development of a variety of calculus and analysis topics, including contributions from diverse cultures.

Knowledge of Discrete Mathematics

Mathematics teachers and teacher candidates specializing in secondary school instruction should be able to apply the fundamental ideas of discrete mathematics in the formulation and solution of problems.

They should be able to demonstrate knowledge of basic elements of discrete mathematics, such as graph theory, recurrence relations, finite difference approaches, linear programming, and combinatorics. They should be able to apply the fundamental ideas of discrete mathematics in the formulation and solution of problems arising from real-world situations, and to use technological tools to solve problems involving the use of discrete structures and the application of algorithms. In particular, they should be familiar with symbolic logic, induction, iteration, and recursion. They should be familiar with applications of discrete mathematics, such as game theory, fair division, and voting schemes. In particular, they should be able to connect discrete mathematics topics with social science applications that occur in the secondary school curriculum. In addition, they should be able to demonstrate knowledge of the historical development of discrete mathematics, including contributions from diverse cultures.

Standard 2: Knowledge of Mathematical Content

Vignette

The first vignette takes place in an elementary number theory class as preservice teachers explore a well-known problem. The problem requires the preservice teachers to draw on mathematical knowledge to uncover unexpected relationships. The instructor's questions provoke reasoning and justification from the preservice teachers as well as additional questions and extensions from them.

2.1—*Unlocking the Locker Problem*

In a university mathematics class for preservice elementary school teachers, students have been studying concepts in number theory. The instructor, Dr. Ong, has posed the following problem, and student groups of four or five are exploring the problem. The instructor has deliberately chosen a problem that will provide opportunities to make connections with concepts explored in number theory.

> One thousand students have lined up in a long hall housing 1,000 closed lockers. One by one the students move through the hall and perform the following ritual: The first student opens every locker. The second student goes to every second locker and closes it. The third student goes to every third locker and changes its state. If it is open, the student closes it; if it is closed, the student

> opens it. The fourth student behaves in a similar manner, changing the state of every fourth locker. Students in line continue in this manner. After all 1000 students have moved down the hall, which lockers are open?

Wanda, Craig, Dina, and Mario have been working together for several minutes, as Dr. Ong circulates among the groups.

Wanda: So the ninth person goes to locker 9 and opens it.

Craig: What about the factors involved?

Dina: Seven is going to change the state of .... Seven stayed open until the seventh person got there. Five stayed open.

Wanda: Those are primes.

Mario: So all primes stay open until the person changes the state. So we know that all primes are closed.

These students are applying ideas and language from their work in number theory. They are communicating with one another and drawing attention to patterns through the use of shared mathematical language. They see making conjectures as a natural part of their work in this class. Dr. Ong approaches the group.

They see making conjectures as a natural part of their work in this class.

Craig: One, 4, and 9 are open.

Dina: These are perfect squares.

Wanda: Let's try 4 squared.

Craig: Just do 16.

Wanda: You can't just do 16, because you might have multiples you have to close or open before 16.

Wanda (turning to Dr. Ong): We're going to conjecture that perfect squares are open. Primes are closed.

Dr. Ong: Why? (When the group offers no explanation, Dr. Ong continues.) You have an interesting conjecture, but why? What is so special about square numbers? What is it about the structure of primes that causes lockers with those numbers to be closed?

Dr. Ong asks questions to help students develop a disposition to question, investigate, and justify. He uses students' discoveries and generalizations to prompt further questions. By modeling this type of questioning, Dr, Ong helps students learn to frame their own mathematical questions and apply their mathematical knowledge to the situation.

Dina: Well, the primes get touched by only that person.

Wanda: But why are the square numbers open?

Dina: Let's look at composite numbers.

Mario: A composite gets hit for each factor.

Craig: Six is 2 and 3, but 4 is 2 and 2, and 9 is 3 and 3. Then why aren't all composites open as well?

The group pursued this problem together for nearly thirty minutes before they arrived at an explanation that satisfied them. Later, when they shared their results with their classmates, they recalled a particular breakthrough.

Wanda: It took us a long time to consider the importance of all factors of each number.

Mario: Yeah, 4 has 1, 2, and 4 as factors. But 6 has 1, 2, 3, and 6 as factors.

Dina: Now it seems so simple. Squares have an odd number of factors, and other composites have an even number of factors.

Craig: The repeated factor means there is one factor instead of a pair of factors.

Yuko had been working in a group that also had found patterns among the squares, but she was not entirely convinced by the discussion up to that point: "I can see how it works for squares like 4, 9, and 25, but how do you know that a very large square number like 576 would be open?"

Yuko's question demonstrates her understanding that inductive reasoning does not constitute a proof. As a result, she has created a need for a convincing argument to support the conjectures made by her classmates.

Other members of Yuko's group added their own questions to extend the problem:

Avery: What if there were more than 1000 lockers?

Hallie: Even if there were only 1000 lockers, what would be the greatest locker number that would be open?

Harvey: What is the greatest number of different people that touch the same locker?

Vignette

This second vignette records a day in a mathematics education class. Preservice secondary school teachers are helped to see the connection between the advanced mathematics they have studied and the school mathematics they will teach.

2.2—*Connecting Content with the Secondary Curriculum: Behavior of Functions*

Dr. McCrone teaches a course on the teaching and learning of secondary school mathematics in a university teacher education program. All her students are enrolled in a teacher development program that leads to a bachelor's degree in mathematics as well as to state certification as secondary school mathematics teachers. At this point in their programs, the preservice teachers have completed several semesters of calculus, linear algebra, geometry, discrete mathematics, statistics and probability, and abstract algebra.

To help her students get a better sense of the scope of the secondary school mathematics curriculum, Dr. McCrone has her students review a variety of current secondary school mathematics textbooks. Some of the textbooks are traditional, single-subject books, such as algebra, geometry, or advanced algebra. Others contain integrated curricula that focus on themes and applications and that develop mathematical strands or topics in the context of problem-based situations. So that her students can place these materials on a continuum of school mathematical concepts, Dr. McCrone also has her students review several middle-grades textbook series, both traditional and integrated.

Several students express surprise that so much sophisticated mathematics content is developed in middle school and secondary school programs. Harriett says, "I didn't learn about exploratory data analysis until I was in college." Others are concerned about how parents would react to textbooks, such as those containing the integrated curricula, that require extensive reading and writing. Still others notice the repetition in some of the textbooks and wonder why so much time is spent relearning concepts that have been included in some of the textbooks several years in a row.

Several students express surprise that so much sophisticated mathematics content is developed in middle school and secondary school programs.

As she is flipping through pages in a second-year-algebra textbook, Robin says, "Hey, I remember this problem. We solved it in calculus."

Dr McCrone asks Robin to write the problem on the board.

> Square corners are cut from an 8 ft. × 10 ft. piece of cardboard. The piece of cardboard is folded to form an open-topped box. What size congruent squares should be cut from the corners in order to form a box with the greatest volume?

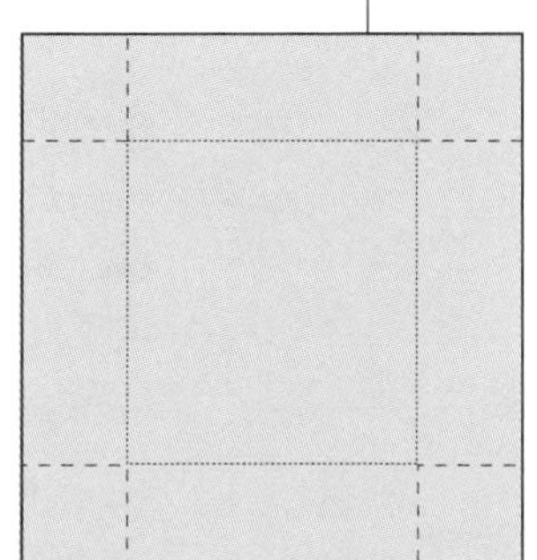

Dr. McCrone asks, "How can students solve a calculus problem in algebra II?" She assigns students to work in their groups to find as many noncalculus methods for solving the problem as possible.

After generating a cubic equation to represent the situation, one group shared a graphical solution in which they used the trace button on a graphing calculator to identify the approximate values of the coordinates of the highest point on the curve. Another group showed how they used a maximum finder on the calculator to identify the maximum value of the graph. A third group illustrated their use of the table function on the calculator to estimate the maximum value.

After the presentations, Robin says, "I guess we'll be responsible for previewing optimization before students get to calculus."

Closing Thoughts
Standard 2: Knowledge of Mathematical Content

The standard of Knowledge of Mathematical Content requires that teachers develop connected, deeply held understanding of the mathematical content and processes described in *Principles and Standards* (NCTM 2000) To help students make connections within mathematics and between mathematics and other disciplines, teachers must be able to make those connections. Teachers must know the value of, purposes for, and connections among various representations to help students learn to flexibly use multiple representations. Likewise, to facilitate student reasoning, problem solving, and communication, teachers must have extensive experience engaging in those mathematical processes. This section has discussed the rationale for strong subject-matter preparation and identified core content for the early childhood, elementary, middle school, and high school levels as prescribed by NCTM through the NCATE standards for the preparation of mathematics teachers.

Standard 3: Knowledge of Students as Learners of Mathematics

The preservice and continuing education of teachers of mathematics should provide multiple perspectives on students as learners of mathematics by developing teachers' knowledge of—

- research on how students learn mathematics;
- the effects of students' age, abilities, interests, special needs, and experience on learning mathematics;
- the influences of students' linguistic, ethnic, cultural, and socioeconomic backgrounds and gender on learning mathematics; and
- ways to affirm and support full participation in, and continued study of, mathematics by all students.

Elaboration

Learning is an active, dynamic, and continuous process that is both an individual and a social experience. Students are naturally inquisitive and have a desire to learn. Their early experiences reflect the excitement of discovery. In school, however, limitations of time, place, and perceptions often constrain their natural curiosity as students encounter environments that are not responsive to them as learners.

The study of general principles of teaching and learning is insufficient for teachers of mathematics because it does not include consideration of the nature of mathematics nor of current research on students' mathematical thinking and its implications for instruction. Students build a variety of perceptions of mathematics as they learn. Some of those perceptions are confused or incomplete; others are remarkably effective. Teachers need opportunities to examine students' thinking about mathematics so that they can select or create tasks that can help students build more valid conceptions of mathematics. Developing multiple perspectives on students as learners of mathematics enables teachers to build an environment in which students can learn mathematics with appropriate support and acceptance.

The study of general principles of teaching and learning is insufficient for teachers of mathematics because it does not include consideration of the nature of mathematics nor of current research on students' mathematical thinking and its implications for instruction.

Research on Students' Mathematical Thinking

Professional development programs, both preservice and in-service, should incorporate current theories and research from mathematics education and the behavioral, cognitive, and social sciences as they relate to mathematics learning. For example, central to current theories is the view of learners as active participants in learning. Learners construct their own meaning by connecting new information and concepts with what they already know, building hierarchies of understanding through the processes of assimilation and accommodation. They learn mathematics when they engage in their own invention and impose their own sense of investigation and structure.

They learn mathematics when they engage in their own invention and impose their own sense of investigation and structure.

The implications of such research and theory building for teaching are continually unfolding as results from research and practice offer new insights and directions for our understanding. Programs for teachers should enable teachers to become active researchers in their own classrooms as well as users and interpreters of research as it relates to their everyday teaching. A sampling of current instructional issues includes—

- the role of language in the development of mathematics understanding,
- the influence of the social environment on the development of shared understandings in the classroom,
- the implications of students' informal mathematical understandings for mathematics teaching and learning,
- the influence of technologies on student understanding,
- the development of students' algebraic reasoning,
- the development of students' statistical and probabilistic reasoning,

- students' perceptions of proof and proving,
- the effect of teachers' mathematical knowledge on student achievement, and
- the interplay of assessment and instruction.

Teachers must be able to interpret research related to instructional issues to determine how they can address those issues in their teaching.

With the help of available technology, the study of students' mathematical thinking can be brought "alive" in new ways. Videotapes can be used to portray developmental sequences in learning or to demonstrate assessing students' developmental levels on the basis of specific learning tasks. Indeed, computer-controlled options allow the development of interactive learning environments that teachers can explore to better understand students' thinking about various mathematical scenarios.

Changed perceptions about what their students can and cannot do affect teachers' attitudes and beliefs about their students and about their teaching strategies.

In addition, such clinical experiences as interviewing students one-on-one or in groups allow teachers at any level to appreciate how much they can learn about students' thinking by talking to students. In conjunction with seminars, courses, or other professional development activities, practicing teachers can learn about current research on students' understanding of mathematics concepts and can validate knowledge of their own students to build deeper understanding of the research and its implications. In follow-up seminars, teachers have opportunities to report and discuss their findings. Changed perceptions about what their students can and cannot do affect teachers' attitudes and beliefs about their students and about their teaching strategies.

The importance of teachers' knowledge of how students learn mathematics cannot be overemphasized. Such knowledge provides direction for the kinds of learning environments that teachers of mathematics create, the tasks they select, and the discourse they foster.

Effects of Students' Age and Ability

Teachers' expectations, founded on knowledge and beliefs about who students are and what they can do, have significant impact on what happens to students in school. Consequently, teachers must understand and appreciate the influences of students' age, abilities (both mental and physical), interests, and experience on students' readiness to learn mathematics.

Teachers must be able simultaneously to perceive mathematics through the minds of their students and to perceive the minds of their students through the mathematics in which they are involved.

How does mathematics appear to an eight-year-old? A twelve-year-old? A fifteen-year-old? Teachers must be able simultaneously to perceive mathematics through the minds of their students and to perceive the minds of their students through the mathematics in which they are involved. Such a perspective requires a thorough knowledge of students' developmental characteristics that emphasizes students' patterns of intellectual, social, and emotional growth. Beyond a general, comprehensive overview, teachers at the middle-grades and secondary school levels need a more detailed understanding of adolescence. Such understandings at all levels must be interwoven with teachers' own developing knowledge about how students learn mathematics.

Teachers' beliefs about students often are tied to their perceptions of students' intellectual abilities. However, research on ability grouping calls into question current tracking practices, as noted by Secada: "Research-based advances in mathematics teaching and learning have shown that the practice of sorting students and educating the few is not necessary, natural, or inevitable" (2000a, p. 1). For example, heterogeneous groups are typically used in elementary schools, and K–5 students generally study the same grade-level content. But at the middle school level, students are often placed into different, ability-ranked mathematics courses (Uecker and Cardell 2000). Although homogeneous grouping in middle and secondary schools is strongly supported by many teachers, no research basis supports this common practice (Slavin 1990; Linchevski and Kutscher 1998). While the debate continues on how ability grouping affects student achievement (Loveless 1998; Kulik 1992), teachers need knowledge about, and experience with, using alternative strategies, such as cooperative and team learning, that work well in heterogeneous environments.

Another problem with tracking in mathematics is that students who have difficulty during their elementary and middle school years with traditional paper-and-pencil computation, both arithmetic and algebraic, are often limited in their access to advanced mathematics. However, computational fluency is not always a valid measure for success at advanced levels of mathematics. Hypothesizing, approximating, estimating, reasoning, problem solving, and communicating are skills and abilities not often tapped or promoted through traditional computational work. In affirming and encouraging full participation by every student, issues surrounding false scope-and-sequence barriers that establish inappropriate prerequisites must be considered in professional development activities with teachers.

Cultural, Linguistic, Gender, and Other Demographic Considerations

Teachers, administrators, and teacher educators, as well as others involved in education, need to examine the very real influence of demographic factors, including culture, language, gender, and membership in a socioeconomic group, on academic performance. The data, as summarized by Berliner (2006), demonstrate the powerful negative effects of poverty on student achievement, even at very young ages. Although the effects may be more subtle, negative treatment on the basis of perceived racial, class, and linguistic differences can and does affect student achievement (NCTM 2005). As students internalize messages they receive from society about their cultural group, they may come to accept negative stereotypes, which in turn affect their performance on tests (Howard 2004–2005).

To close the achievement gap, teachers must be required to confront their own beliefs about, and expectations for, students from diverse cultural groups. They must also learn

To close the achievement gap, teachers must be required to confront their own beliefs about, and expectations for, students from diverse cultural groups.

to understand their students' backgrounds so as to implement "curricula that are culturally relevant and methods of instruction that are culturally sensitive" (NCTM 2005). Recognizing and enfolding mathematical aspects of ethnic and cultural identity into lessons helps to provide an impetus for further study of mathematics. Students from every background can benefit from learning how diverse cultures, in particular those of underserved and underrepresented groups, have developed number, measurement, geometry, and other mathematical concepts and applied them in their daily living. To be prepared to lead culturally rich mathematical discussions, teachers themselves need knowledge of those cultural achievements and models for how to incorporate them into their teaching.

Teachers must also understand the importance of everyday context as it relates to students' interest and experience. Instruction should incorporate real-world contexts and students' experiences and, when possible, should use students' language, viewpoints, and culture. Students need to learn how mathematics applies to everyday life and how mathematics relates to other curriculum areas as well.

Language and its role in students' understanding and doing mathematics need attention in programs for the development of teachers. In some circumstances, students' understanding of the language used to communicate mathematics may be incomplete or incorrect; those misunderstandings can create subtle or overt barriers to success in the mathematics classroom. Although students may lack appropriate vocabulary and syntax to express themselves mathematically, they still may be able to learn and demonstrate sophisticated knowledge of mathematics. Teachers' knowledge of their students' cultural backgrounds and the implications of that knowledge for their teaching are crucial in recognizing the impact of language on learning.

Improved mathematics instruction "is grounded in the process of redefining teaching as creating learning environments that capitalize on students' home language and experiences [instead of] ignoring and devaluing them."

All children who are English language learners should be taught mathematics in their first language until they have attained English proficiency. However, as Khisty (1995) demonstrated in her research with Hispanic students, a successful approach involves more than which language is used in the classroom; it involves the questions that are asked, the answers that are accepted and probed—in general, the nature of the teacher's use of language when teaching mathematics. She argued that improved mathematics instruction "is grounded in the process of redefining teaching as creating learning environments that capitalize on students' home language and experiences [instead of] ignoring and devaluing them" (Khisty 1995, p. 295).

Attention continues to be given to gender differences in learning styles and academic behaviors. Data collected from such sources as the National Assessment of Educational Progress (NAEP) and SAT Tests show that although boys still score somewhat higher than girls in mathematics, considerable progress has been made in addressing the "gender gap" in precollege mathematics performance and participation (Jacobs and Becker 2001). Yet some sizeable discrepancies remain, particularly in earned degrees and career choices in mathematics-related fields (Fox and Soller 2001). Hypotheses proposed for those differences have included girls' lack of self-confidence in their mathematical abilities, their perception of mathematics as a male domain, and a belief that mathematics is unrelated to everyday concerns. Earlier considerations of the gender gap have focused on how to change girls' perception of, and involvement in, mathematics. However, current work indicates that females make sense of information and learn in ways that are significantly different from those assumed in the traditional approach to teaching mathematics. Programs for mathematics teachers need to provide access to the literature that explicates gender-based learning differences and identifies motivational strategies to engage both boys and girls in the study of mathematics.

Supporting Students and Encouraging Participation

Promoting equity in the classroom must be a priority for all teachers. Although perhaps subtly and unintentionally, students are not always treated equitably. Determining whether inappropriate differential treatment of students exists is one way to begin addressing the problem of reaching all students. Are students treated differently on the basis of their gender or their cultural or linguistic backgrounds? Teachers need help in learning to monitor classroom interactions; a colleague observing or videotaping a class can be of assistance in this endeavor. Recording instances of positive and negative feedback and of disciplinary and social interactions, as well as the name of each student who does and does not receive attention, can foster insights into unconsciously biased behaviors. If inequities are identified, then strategies need to be developed to help a teacher address them. Such strategies should be discussed in professional development activities for teachers.

Grouping of students, classroom climate, choice of materials, topics, activities, assessment, and instructional strategies all have an effect on students' participation in, and attitude toward, learning mathematics. Students' age, abilities, and interests as well as the academic, ethnic, linguistic, cultural, and gender makeup of the class should be considered when planning activities that promote a positive and challenging learning environment for all students. A genuine respect for, and understanding of, students as individuals and as participants in a community of learning are essential to promoting the kinds of experiences that involve all students in mathematics.

A genuine respect for, and understanding of, students as individuals and as participants in a community of learning are essential to promoting the kinds of experiences that involve all students in mathematics.

Standard 3: Knowledge of Students as Learners of Mathematics

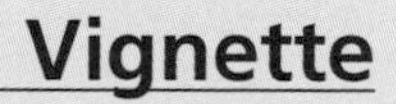

In the first vignette, preservice teachers learn to apply research about student understanding to classroom situations. The preservice teachers analyze students' responses to problems designed to reveal and develop students' understanding of fractions. In particular, they struggle to see the mathematical concepts from the students' perspective.

3.1—*Assessing and Building on Students' Fraction Concepts*

In her mathematics methods class, Dr. Palmer has been trying to help prospective

Observing and interviewing students can help teachers revise their assumptions about how students learn mathematics and learn to interpret students' words, representations, and ways of communicating their mathematical ideas.

elementary school teachers learn to "listen mathematically" to students. Observing and interviewing students can help teachers revise their assumptions about how students learn mathematics and learn to interpret students' words, representations, and ways of communicating their mathematical ideas. They have been reading case studies of young students, watching videotapes, and reading theoretical pieces on how students learn mathematics. This week, Dr. Palmer assigned students the task of closely examining some aspect of students' understanding in their field classrooms. This afternoon she is meeting with Mr. Konook and five prospective teachers who work in his fourth-grade class to discuss their observations.

Damon immediately brings up a conversation he had with one of the students that afternoon. "The last problem on the board asked students to show which was more, 2/6 or 1/3. Tia wrote

$$2/6 > 1/3.$$

When I asked her to explain, she drew this picture:

Because she did not draw each piece the same size, her picture of 2/6 was indeed larger than 1/3."

"That's interesting," Lisa said. "Latalya got the same answer but for a different reason. She drew this picture:

and said that 2/6 is more than 1/3 because 1/3 is the same as one out of three and 2/6 is the same as two out of six. We have three dogs, one of them is black. We have six hamsters, two of them are black. We have more black hamsters than black dogs."

Going beyond their written answers, they learn to probe the depth of students' understanding.

The classroom observations show the variety of strategies that students employ to make sense of fractions. Through observations, questioning, and listening to students' explanations, the prospective teachers uncover the thinking underlying their approaches. Going beyond their written answers, they learn to probe the depth of students' understanding.

"Wait," Maura said, "I don't understand. One of three dogs and two of six hamsters are both 1/3."

But she is right," Peter exclaimed, "because she does have more black hamsters than black dogs."

Lisa explained, "I wouldn't have understood how Latalya got her answer if I hadn't asked. I just assumed that she didn't understand the problem. If that were a question on a test, her answer would not reflect what she does know about fractions."

"But we know that 1/3 is equivalent to 2/6 when we are talking about two equal-sized things. Don't we want them to see that? Isn't this just confusing them?" asked Maura.

Kamisha added, "If Latalya has twenty cats and two of them are black, should she say that 2/20 was the same amount as 2/6?"

"What do you think she would do?" Dr. Palmer asked Kamisha.

"It seems like she is only using the numerators to determine which of them is more."

"What difference does that make?" asked Dr. Palmer.

"She's not thinking about how much of the denominator it is."

"What do you mean?" Lisa asked. "She seems to understand that the denominator is the number of things in thc set."

"But when she is comparing two fractions of different-sized sets, she is only considering the number of parts and not *how much of the total set* they are."

"That is one of the things that makes fractions complex," Dr. Palmer commented. "Not only are we interested in the numerator and denominator, but we also want to know about their relationship to [each other]."

"Perhaps she is thinking about that, but she is considering two different-sized sets," Lisa suggested.

"One of the things Latalya is comparing is twice as big as the other, and all the pieces are the same size. So two of the larger set is more than one of the smaller set." Lisa went to the board and drew the boxes around the circles in Latalya's drawing to show that she set of hamsters was twice as large as the set of dogs.

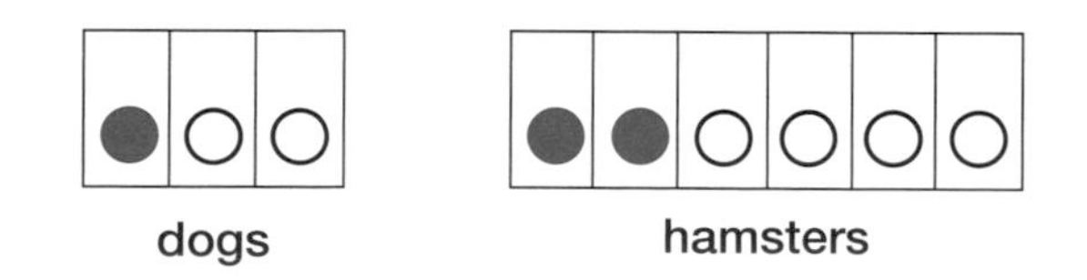

"Her 2/6 is more than her 1/3," she concluded.

"Tia is doing something different," Peter volunteered. "She is comparing two equal-sized sets. She has just divided them up unequally. She may not realize that each thing should be divided into equal pieces, although she seems to understand the part-whole relationship."

Turning to Mr. Konook, Damon suggested, "Maybe the task should have already had the pictures drawn with it; then these problems wouldn't have come up."

"Actually," explained Mr. Konook, "I intentionally put the problem on the board as a symbolic statement and asked them to show how they got their answers. I wanted to see what they understood about comparing fractions. Listening to your descriptions of the conversations you had with these students has been very helpful."

"Wow," exclaimed Damon, "I was thinking that you would want to give the students a model to use to help them get the answer and avoid the confusion. But it seems like having them come up with their own models and allowing for some confusion revealed what the students do and don't understand. These two students got the same answer, but for very different reasons."

The foregoing exchange shows professional collegiality in analyzing and evaluating effective teaching practices. Through that interaction, the preservice teachers had an opportunity to see how children's informal understandings can be accessed through teacher questioning. By analyzing the students' responses, both the preservice teachers and the practicing teacher were able to think about how to draw on students' existing conceptual models and design or refine tasks to help them move toward more flexible understandings.

Vignette

In the next vignette, a team of middle-grades mathematics teachers works together to explore one sixth-grade student's thinking through a videotaped interview. The team members consider the effects of age on the difficulty level of the concepts involved in the problem.

3.2—*Exploring Statistical Thinking: An Interview with Sara*

Dr. Williams, the district mathematics coordinator, and several grades 6–8 teachers meet monthly to explore ways to improve their teaching practices. They have been reading about and discussing ways to investigate their students' thinking. In the course of one such discussion, they began to explore the issue of teaching statistics—particularly the concept of mean. Dr. Williams has brought in a videotape of an interview with a sixth-grade student.

Dr. Williams: Before we look at this tape, I want to pose a problem. You should solve the problem and pay particular attention to the strategy you use.

You have nine bags of different kinds of potato chips, and you know that the average cost for a bag of chips is $1.38. What might be the actual prices of each of the nine bags of chips?

The teachers work individually and then compare strategies with one another. Then they spend time discussing their various strategies. Following those conversations, Dr. Williams plays the videotape, which leads to a discussion and analysis of the strategy shown in the tape.

Mary: I noticed that Sara, the student in the tape, seemed comfortable with the problem. It made some sense to her because she is familiar with the cost of bags of potato chips.

Irv: Yes, right away she said, "Well, I know they can't all be the same price. That's not the way it is in the store." Unlike her, I immediately decided to do a "quick and easy" solution and make each bag the same price—but that certainly isn't real!

Another teacher went further and began to consider what Sara actually understood. The teachers realized that "average" may be a more complex concept than she originally imagined.

Natasha: I was intrigued when she started to figure out actual prices. She knew she wanted some of the prices to be less than $1.38 and some to be more. She also was quite emphatic about stating that there had to be at least one bag with a price of $1.38. Earlier Jody showed us that we could actually have nine bags of potato chips with an average price of $1.38, and none of the bags actually had this price. Sara either wasn't comfortable with this as a possibility, or it may not have even occurred to her. I bet this may be a difficult concept to understand for students of this age.

Dr. Williams: What did you notice about the way the interviewer worked with Sara?

Don: Well, he posed the problem and explained the materials that were available, pointing out the pictures of the bags and the markers she could use to record prices. Then he became an observer. Occasionally he asked a question for clarification, but he really didn't involve himself in her work. Gee, it would be difficult for me to know when to intervene and when to keep silent. I would want to jump in with suggestions!

The teachers and Dr. Williams continue the discussion, commenting on the student's response to the problem. Finally, Dr. Williams presents a synthesis from the particular research project, detailing other problems used in the research and providing an overview of the results of the research, which focused on the developmental differences that were found among the fourth-, sixth-, and eighth-grade students who were interviewed.

The teachers agree to pose the same problem to some of their students and bring the results to the next study group meeting. They are intrigued by the variation in student performance by age and wonder whether they will get similar results in their own classrooms.

Vignette

The following vignette gives a snapshot of student teachers' conversation as they confront the complex issues of (1) managing classrooms without introducing gender bias and (2) meeting the needs of second language learners without neglecting other students. Although they reach no conclusion, their recognition of the issues and formulation of their points of view are necessary parts of the learning process.

3.3—*Facing a Teaching Dilemma: Gender Interaction*

The student teachers in Dr. Dreyfus's seminar group are discussing a teaching dilemma described in an article on how teachers manage to teach. In the article, a teacher explores the behavior of fifth graders in male-female interactions:

> The students in my classroom seem to be allergic to their peers of the opposite sex. Girls rarely choose to be anywhere near a boy, and the boys actively reject the girls whenever possible. This has meant that the boys sit together at the table near one of the blackboards and the girls at the table near the other.
>
> The fifth-grade boys are particularly enthusiastic and boisterous. They engage in discussions of math problems with the same intensity they bring to football. They are talented and work productively under close supervision, but if left to their own devices, their behavior deteriorates and they bully one another, tell loud and silly jokes, and fool around with the mathematics materials. Without making an obvious response to their misbehavior, I developed a habit of routinely curtailing these distractions from the lesson by teaching at the blackboard on the boys' end of the classroom. This enabled me to address the problem of maintaining classroom order by my physical presence. But my presence near the boys had inadvertently put the girls in "the back" of the room. One of the more outspoken girls impatiently pointed out that she had been trying to get my attention and thought I was ignoring her. She made me aware that my problem-solving strategy, devised to keep the boys' attention, has caused another, quite different problem. The boys could see and hear more easily than the girls, and I noticed their questions more readily. Now what was to be done?
>
> I felt that I faced a forced choice between equally undesirable alternatives. If I continued to use the blackboard near the boys, I might be less aware of, and less encouraging toward, the more well-behaved girls. Yet, if I switched my position to the blackboard on the girls' side of the room, I would be less able to help the boys focus on their work. Whether I chose to promote classroom

order or equal opportunity, it seems that either the boys or the girls would miss something I wanted them to learn.

Dr. Dreyfus opens the discussion of this complex issue by commenting, "Dilemmas are an inherent part of teaching. We may 'know' what is desirable in terms of theory, research, and expected practice, but often the route we take to achieve one goal is at odds with another. What are some of this teacher's alternatives?"

Ian immediately responds, "You know, I have a similar situation in my seventh-grade class. The girls are generally more attentive and less distractible, so I spend more time focused on the boys. I tried mixing up the boys and girls, assigning seats that distributed them around the room. Now they spend more time talking with each other than paying attention to mathematics. I don't feel I have solved the problem."

"It's hard to be attentive to all the needs of my students," notes Kadisha, who is working in a sixth-grade ESL class. "I'm particularly aware of the students who have language difficulties. I give them a lot more attention than other students. Maybe the teacher in the article should try some small-group activities. Then she could walk around the room and spread out her attention."

"But whole-class discussions are a central part of my mathematics classes," Maureen responds. "By the time the students are in ninth grade, the pattern has been established. The boys dominate the discussion, and many girls are reluctant to contribute. I find myself asking the girls less challenging questions or changing the tone of my voice in order to encourage them to participate. So in a sense I'm countering one imbalanced situation with another. But I'm hoping to gradually change that."

"I don't think we can expect changes overnight," Rika reminds her peers.

Issues of promoting equity must be addressed in light of other pedagogical concerns, such as the role of discourse. The preservice teachers described here are gaining a sense of the complexity of teaching and of ways in which they need to monitor their own behaviors in the classroom. They are developing sensitivity to all students in their classrooms and seeking ways to manage the ongoing dilemmas of teaching.

The preservice teachers are gaining a sense of the complexity of teaching and of ways in which they need to monitor their own behaviors in the classroom.

Closing Thoughts
Standard 3: Knowledge of Students as Learners of Mathematics

Knowing students as learners of mathematics requires knowledge of learning theory as well as knowledge of the effects of individual characteristics, the influences of gender and sociocultural background, and ways to encourage participation in the learning process by all students. This Standard acknowledges the influence of those factors on mathematics learning and challenges teachers to work with colleagues and others to find ways to incorporate an understanding of those factors into their professional decision making.

Standard 4: Knowledge of Mathematical Pedagogy

The preservice and continuing education of teachers of mathematics should develop teachers' knowledge of and ability to develop or select, implement, and reflect on—

- instructional materials and resources, including appropriate technology;
- ways to represent mathematics concepts and procedures;
- instructional strategies and classroom organizational models;
- ways to promote discourse and foster a sense of mathematical community; and
- means for assessing student understanding of mathematics.

Elaboration

Mathematical pedagogy, sometimes referred to as pedagogical content knowledge, focuses on the ways in which teachers help students come to understand and be able to do and use mathematics. This Standard identifies several components of content-specific pedagogy that are essential to high-quality teaching. These components act as a series of lenses through which teachers filter their knowledge of mathematics and of students to enrich and enhance the teaching of mathematics.

Teachers need a well-developed framework for identifying and assessing instructional materials and technological tools, and for learning to use those resources effectively in their instruction.

Teachers are responsible for posing worthwhile mathematical tasks to actively engage students in mathematical thinking and so that students have opportunities to see connections within and across disciplines. Teachers may choose already developed tasks or may construct tasks to focus students' mathematical learning. To do so, they often rely on a variety of instructional materials and resources, including physical materials, textbooks, online resources, computer software, calculators, and materials shared among colleagues. Teachers need a well-developed framework for identifying and assessing instructional materials and technological tools, and for learning to use those resources effectively in their instruction. Such a framework is built from a teacher's understanding of mathematics and of what constitutes worthwhile mathematical tasks as well as knowledge of ways to represent mathematical ideas.

Representations serve as vehicles for examining mathematical ideas.

Modeling mathematical ideas through the use of representations (physical, visual, graphical, symbolic) is central to the teaching of mathematics. Representations serve as vehicles for examining mathematical ideas. As stated in the National Research Council's (2001) review and synthesis of research on mathematics learning,

> Mathematics requires representations. In fact, because of the abstract nature of mathematics, people have access to mathematical ideas only through the representation of those ideas. Although on its surface school mathematics may seem to be about facts and procedures, much of the real intellectual work in mathematics concerns the interpretations of mathematical ideas.
>
> (National Research Council 2001, pp. 94–95)

Teachers need a rich, deep knowledge of the variety of ways mathematical concepts and procedures can be modeled, understanding both the mathematical and developmental advantages and disadvantages in making selections among the various models.

Not only do teachers need to be familiar with a variety of representations, they must be comfortable with helping students construct their own representations. Designing instruction involves a variety of decisions about the role and use of representations: Should structured representational materials be used? If so, which representational materials furnish the most appropriate model for helping develop the concept at the given point in instruction? If not, how can students refine the existing models they have been using or develop new models for themselves? Of the various options, which representations are most transparent for illustrating particular concepts so that students have the best opportunities for making sense of the mathematics? Numerous other questions surface before instructional choices are complete. Choosing, modifying, or constructing representations are central pedagogical considerations that must be continually addressed.

Mathematics instruction is often approached in terms of stating and exemplifying rules—the "tell, show, and follow my lead" model. That approach is based on the assumption that information can be presented by telling and that understanding will result from being told. However, such an instructional approach does not work for many students. It does not incorporate two crucial developmental components: the process of assimilation and the issue of readiness. Essentially, in the telling approach, students are "ready" intellectually when the teacher is ready for them to receive the information. Learning through such an approach often fails to promote transfer of mathematical information to new situations.

Teachers, both preservice and in-service, need to be exposed to alternative forms of instruction that permit students to build a repertoire of mathematical knowledge and to develop abilities for posing, constructing, exploring, solving, and justifying mathematical problems and concepts (NCTM 2000). Promising models for such instruction are all highly interactive. In such models, teachers both model and elicit mathematical discourse by asking questions, following leads, and conjecturing rather than presenting faultless products.

Teachers need to be taught how to create learning environments that encourage students' questions and deliberations—environments in which the students and teacher are engaged with one another's thinking and function as members of a mathematical community. In such a community, the teacher-student and student-student interactions give teachers opportunities to make diagnoses, offer guidance, and model mathematical thinking while giving students opportunities to challenge others' claims and defend their own conjectures.

Teachers need to be taught how to create learning environments that encourage students' questions and deliberations—environments in which the students and teacher are engaged with one another's thinking and function as members of a mathematical community.

Teachers need to learn strategies that will help them promote the participation essential to engaging students in mathematics. Increasing the amount of time students spend working together supports the development of discourse and community.

Working in groups, students gradually internalize the discourse that occurs, challenging themselves by asking for reasons and, in general, accounting for their own mental work. Another practice that supports students' participation involves shifting responsibility from teacher to student for control of learning by expecting students to make commitments to their answers. Further, students' reflective processes can be developed by focusing efforts on interpreting problems, describing strategies for solutions, and justifying and defending the results.

Teachers' willingness to be flexible and curious about mathematics with their students is central to their ability to promote mathematical discourse.

Teachers' willingness to be flexible and curious about mathematics with their students is central to their ability to promote mathematical discourse. Engaging in personal dialog with colleagues about mathematics and mathematics instruction and establishing a classroom that encourages engagement in discourse help teachers deepen, extend, and enhance their knowledge of mathematics and of their students' knowledge of mathematics. Therefore, during teacher preparation and ongoing professional growth, teachers need to experience such mathematical conversation in their own coursework, in mathematics classes as well as pedagogy classes, and reflect on how to promote such conversation among their own students.

Strategies and practices need to be implemented that enable teachers and others to assess students' mathematical proficiency in a manner that emphasizes important mathematics and is consistent with the way students learned it. Assessment should not focus primarily on rank ordering students according to their levels of achievement. Instead, we need to move toward an approach to assessment that is philosophically consistent with the vision of teaching and learning described throughout this and other *Standards* documents (NCTM 1989, 1991, 1995, 2000).

Assessment should be an integral part of mathematics teaching. Through assessment, teachers learn how students think about mathematics and what they are able to accomplish. Moreover, students obtain feedback so as to resolve inconsistencies and deepen their understandings of mathematics.

Professional growth experiences, therefore, should supply teachers with assessment strategies that focus on addressing students' understanding of mathematical concepts and procedures and the relationships among them, and their abilities to reason mathematically and apply their knowledge to a variety of problem situations. Teachers need to learn how to align assessment with instructional goals and to consider the purposes of assessment as they select or develop the means of assessment. In addition, teachers need to understand the issues surrounding assessment in general, the arguments related to those issues, the distinctions between classroom assessment and accountability testing, and proposed alternatives for unifying instruction and assessment.

Furthermore, teachers need to learn how to analyze, communicate, and act on the results of both formal and informal assessments. The communication may involve only a classroom teacher using information collected to provide direction for working with an individual student, group of students, or class of students in continuing instruction. Teachers also need to communicate with one another about learning and teaching.

Results from assessment are often the catalyst needed for jointly diagnosing students' understandings and misunderstandings, designing curriculum, planning instruction, or initiating further assessment efforts. Finally, teachers need to be introduced to various ways to assess students and to communicate the results in a more formal way to students, parents, and others in a school district to indicate students' understanding of mathematics.

The aspects identified in this Standard as "mathematics content pedagogy" are integral to the effective teaching of mathematics. Teachers' knowledge and their ability to use and evaluate those components develop over time. Decisions about instructional materials are intimately associated with decisions about ways to represent mathematics concepts and procedures. Choices for instructional strategies and classroom organizational models both evolve from and influence such decisions. Finally, the discourse of the classroom and the ongoing assessment are part of a dynamic interaction that results in the further development of teachers' pedagogical content knowledge.

Standard 4: Knowledge of Mathematical Pedagogy

Vignette

In the first vignette, Pre-K–12 teachers are introduced to assessment strategies at a professional development conference. The interview, an alternative or supplement to paper-and-pencil methods, gives teachers a different window on student thinking. In addition, the experience of sharing their thinking aloud helps students be better prepared to contribute to classroom discourse.

4.1—*Considering Assessment Strategies: The Interview*

A two-day conference on assessment is being held by a regional NCTM affiliate. Teachers are being asked to think very differently about both the purposes of and ways to conduct assessment. Teachers from K–12 schools and early childhood learning centers have come to learn how to design comprehensive pre-K–12 assessment programs that go beyond classroom-level or large-scale tests. The conference comprises a number of sessions, each demonstrating an alternative form of assessing what students know and understand about mathematics.

Presenter: Assessment is often thought of as being synonymous with paper-and-pencil testing. If paper-and-pencil tests are the only strategy for gathering information, teachers may know very little about their students' understandings of mathematics. We would like to broaden this conception to include a multitude of ways of determining what a child actually understands.

We would like to broaden this conception to include a multitude of ways of determining what a child actually understands.

The presenter first shows the audience a subtraction paper completed by Myra, a first-grade student. On the paper are twenty-five exercises dealing with subtracting a single digit number from a number in the range 5–19.

Four exercises on Myra's paper are circled as having incorrect answers. The audience is asked to hypothesize about what might be Myra's difficulties. After some discussion, the participants conclude that the errors are inconsistent, and most agree that Myra probably just got bored.

Next they view a videotape of a three-minute interview with Myra in her classroom. In the tape, Myra responds to an open-ended task that allows her flexibility in interpreting the task and demonstrating her understanding. Myra is presented with a large number of cubes (in the taped episode, thirty-six) and is asked how many cubes are present. We see Myra counting as she points to individual cubes:

Myra: 1, 2, 3, 4, ..., 26, 27, 28, 97, 96, 95, 98, 99. Yes, there are 99 cubes.

Interviewer: Myra, you remember how we grouped cubes by making sets of ten? (Myra nods enthusiastically.) Do you think you could do that for me right now?

Myra begins to put cubes together. The tape zooms in as Myra is counting the remaining single cubes, having made three sets of ten cubes.

Myra: One, ... two, ... three, ... four, ... five, ... six. Six. (Myra looks up and smiles triumphantly.)

Interviewer: That's very nice. Now, can you tell me how many you have all together?

The interviewer deliberately maintains a supportive and encouraging attitude because the goal is to find out what the child understands and not to intervene.

In this episode, the interviewer deliberately maintains a supportive and encouraging attitude because the goal is to find out what the child understands and not to intervene.

Myra wrinkles her forehead in concentration.

Myra: (Smiling). There are nine all together.

Some of the teachers in the audience react to the tape.

Lauren: Look! She counted the six single cubes and the three sets of ten cubes, which results in her answer of nine. I don't think she has a sense of the structure of base-ten place value.

Ed: Yes, the fact that she counted in this way, even after she had actually put them together herself, makes me think she doesn't understand grouping.

Anita: And from looking at her paper, we thought she was doing fine. Taking the time to talk and ask students to explain their answers and to watch them work in this form of mini-interview is essential to knowing where students are. It is a technique that can be used at all levels. I would like the teachers at my school to see this video and learn some ways to interview their students.

During the session just described, the teachers learned that paper-and-pencil tests are limited in the kinds of information they can provide about students' thinking, especially for young children or children who have limited abilities to communicate in writing. The information obtained by the interviewer clearly identified a need for an instructional intervention and raised questions about how many other misunderstandings may have gone unrecognized.

Vignette

In the following vignette, preservice teachers learn about the variety of ways students can represent a problem. They also consider how to foster mathematical discourse that highlights multiple representations and solutions in the classroom.

4.2—*Learning to Recognize and Exploit Multiple Solution Methods*

In his course for preservice teachers on the teaching and learning of secondary school mathematics, Dr. Glendon emphasizes mathematical problem solving. He includes opportunities for preservice teachers to solve nonroutine problems, to discuss problem-solving strategies, to consider ways of helping students become better problem solvers, and to assess secondary school students' problem-solving performance.

As a culminating activity the preservice teachers have been assigned the task of choosing a rich problem and using it in an interview with a student or pair of students during a prepracticum visit to a secondary school. They are to prepare a written report on the students' methods of solving the problem and what those methods reveal about their understanding of the mathematics.

Three of the preservice teachers have chosen to interview tenth-grade students using the following problem.

> A textbook is opened at random. The product of the numbers of the facing pages is 3192. To what pages is the book opened?

This problem has the potential to shed light on secondary school students' understanding of several different mathematical ideas, their problem-solving processes, and their ability to make connections.

The three preservice teachers, Michelle Tremblay, Peter Marshall, and Ruth Wong, have decided to make a joint report on their findings.

Michelle reports that the student she interviewed at first seemed confused by the problem and reluctant or unable to get started with it. Michelle asked the student, "Can you explain in your own words what the problem is telling you?" Later, she asked, "What operation was done with the two page numbers to get the answer 3192?" Michelle comments, "The questions I asked weren't really hints; they were just the encouragement the students needed to get into it."

In the page-numbers activity, the preservice teachers consider questioning techniques. They commend Michelle on her choice of questions that enabled her student to get beyond a roadblock without telling the student what to do. Michelle, Peter, and Ruth talk about how difficult it was to ask questions that would encourage students to share their thinking and their solution strategies.

Peter begins the discussion of students' strategies for solving the problem by describing the guess-and-test approach taken by the pair of students he interviewed. "Darlene always guessed two numbers that are consecutive (like 81 and 82, 46 and 47), but Jill made wild guesses including pairs of numbers that were not even consecutive (like 31 and 100, 62 and 75). After Darlene and Jill had each worked separately for a couple of minutes, I asked them to compare their results so far," Peter reports. "I didn't even have to tell them what kind of guesses were better. Right away Jill realized that her nonconsecutive pairs weren't helpful, and she was able to suggest the next guess—and that turned out to be very close."

Michelle reports that about half the students interviewed used an algebraic approach; some did so on their own initiative, whereas others did so following a general hint. "They didn't have any trouble deciding to let x be one page number and $(x + 1)$ be the other, and they all got $x(x + 1) = 3192$," she says.

"But from there, different students had different problems. Beth wrote

$$x(x + 1) = 3192$$

$$(x + 1) = 3192/x$$

but didn't know how to go on from there.

Bruce wrote

$$x\text{^}2 + x = 3192$$

$$x\text{^}2 + x - 3192 = 0$$

$$(x + \_\ )\ (x - \_\ ) = 0$$

but got stuck when he tried to factor the quadratic. I think he just gave up because 3192 was such a large number. The coefficients in the quadratics he had seen in class were always much smaller numbers."

Ruth notes that most of the tenth-grade students interviewed by the preservice teachers used calculators as they tried to solve the problem by a guess-and-test approach, which her student, Marc, referred to as "the long way." However, none of them thought of the square-root operation. "I was a little disappointed that no one realized that would be a good estimate for the page numbers," she says. "As a follow-up question, after Marc had found the correct page numbers by trial-and-error, I asked, 'If you had known the value of the square root of 3192, would that have helped you to solve the problem?' But he didn't seem to see the connection I was hinting at."

Peter adds to the discussion of alternative solution methods. "Michelle, Ruth, and I thought that students might use a factorization approach in solving this problem, but none of them did. We expected that the students might try to break down 3192 into its prime factors and then reassemble the pieces to make two factors that are consecutive numbers."

The other preservice teachers compliment Michelle, Peter, and Ruth on their abilities to get students to share their solutions and to assess the students' understanding of the concepts associated with the problem. They also wonder how teachers can adapt the interview techniques that Michelle, Peter, and Ruth used to get students to share their strategies to a full-class setting. Dr. Glendon asks the preservice teachers for their ideas on how to facilitate class discussions that encourage students to share solutions and assess one another's solution methods in a supportive environment.

Closing Thoughts
Standard 4: Knowledge of Mathematical Pedagogy

This Standard addresses a type of specialized knowledge that a mathematics teacher needs beyond that required of other teachers. Mathematics teachers need to know models, strategies, and tasks that will best aid students in constructing understanding of concepts and developing computational fluency. Mathematics teachers need to have the tools to probe students' thinking and interpret students' responses to mathematical tasks. Mathematics teachers need to know how to create a classroom community in which all members have the responsibility to appeal to mathematical reasoning so as to assess the quality of one another's arguments. That knowledge can begin to be developed in teacher preparation programs, but it must also continue to grow by reflection on classroom experiences throughout each teacher's professional career.

Mathematics teachers need to know how to create a classroom community in which all members have the responsibility to appeal to mathematical reasoning so as to assess the quality of one another's arguments.

Standard 5: Participation in Career-Long Professional Growth

The preservice and continuing education of teachers of mathematics should provide them with opportunities to—

- examine and revise their assumptions about the nature of mathematics, how it should be taught, and how students learn mathematics;
- observe and analyze a range of approaches to mathematics teaching and learning, focusing on the tasks, discourse, environment, and assessment;
- work with a diverse range of students individually, in small groups, and in

large-class settings with guidance from, and in collaboration with, other mathematics education professionals;

- analyze and evaluate the appropriateness and effectiveness of their teaching, reflecting on learning and teaching individually and with colleagues;
- participate actively in the professional community of mathematics educators;
- engage in proposing, designing, and evaluating programs for professional development specific to mathematics; and
- advocate in school, community, and political efforts to effect positive change in mathematics education.

Elaboration

The practice of teaching, the growing sense of self as a teacher, and the continual inquisitiveness about new and better ways to teach and learn all serve teachers in their quest to understand and change the practice of teaching.

This Standard addresses issues that are at the heart of teacher development. The goal of teacher education is to "light the path" for those who follow, providing directions on how to plan and teach mathematics. In the career-long process that is teacher education, teachers should be enabled "to acquire and regularly update the content knowledge and pedagogical tools needed to teach in ways that enhance student learning and achievement" through structures that allow teachers to "grow individually in their profession and to contribute to the further enhancement of both teaching and their disciplines" (National Research Council 2000, p. 10). The practice of teaching, the growing sense of self as a teacher, and the continual inquisitiveness about new and better ways to teach and learn all serve teachers in their quest to understand and change the practice of teaching. This standard considers both the development of mathematics teachers and their ongoing growth as professionals.

Development of Mathematics Teachers

The nature and kinds of teaching experiences that should be part of the preservice and continuing education of teachers of mathematics are varied and numerous. For preservice teachers, such experiences involve opportunities to work one-on-one or with small groups of students in clinical settings that permit them to assess, tutor, or engage in microteaching. They need a sequenced program that gives them opportunities to be in classroom settings for a variety of purposes and with increasing levels of responsibility. Finally, they need long-term placements that permit them to become the teachers of students under the guidance and support of both a cooperating practitioner and another mathematics educator, usually a supervisor from a university or college.

During the first few years, teaching is an intensely focused experience that centers on the students for whom the teacher is responsible and on the teacher's growing sense of self as a teacher of mathematics. Colleagues and supervisors can function informally and formally as resources during this time of transition between the sometimes-overwhelming responsibilities of teaching various courses and grade levels for the first time and the confidence that is acquired through a few years of successful teaching experience. Indeed, beginning teachers often welcome and seek the advice of more experienced teachers to obtain guidance and observe some diversity in models of how to teach.

Experienced teachers have different needs. They have a general frame that surrounds their picture of teaching, and they understand the ebb and flow of the learning process as it proceeds daily, weekly, and monthly throughout the school year. They are better able to anticipate timing, overall organization and management, and student responses. Their repertoire of instructional methods has "filled out," and they often can successfully anticipate what will work or not work in the classroom. Nevertheless, they, too, can benefit from the trusted guidance of colleagues and supervisors to assist them in assessing their teaching and making improvements. In addition, when teaching new material or trying out new methods of teaching, teachers are also in a position to assess what works well for them and their students.

Good mathematics teaching is enhanced by conversations with colleagues and supervisors who know mathematics and have been successful in teaching mathematics. Preservice teachers should have opportunities to teach with exemplary mathematics teachers. They should be supervised by teacher education faculty who know mathematics and are experienced mathematics teachers themselves. Practicing teachers also should involve colleagues or teacher educators with backgrounds in mathematics teaching when they are exploring new ways to teach or seeking feedback on current teaching strategies. Mathematics has its own content and pedagogy. Only those knowledgeable about the associated special issues and experienced in the field should serve as mentors or supervise teachers' clinical and field-based learning experiences.

Good mathematics teaching is enhanced by conversations with colleagues and supervisors who know mathematics and have been successful in teaching mathematics.

Essentially, being a teacher of mathematics means developing a sense of self as a teacher of the subject. Such an identity grows over time. It is built from many different experiences with teaching and learning. Further, it is reinforced by feedback from students that indicates they are learning mathematics, from colleagues who demonstrate professional respect and acceptance, and from a variety of external sources that demonstrate recognition of teaching as a valued profession. Confident teachers of mathematics exhibit flexibility and comfort with mathematical knowledge and commitment to their own professional growth within the larger community of mathematics educators.

Teachers' Continuing Growth as Professionals

Teaching mathematics is a challenging profession and, as such, demands ongoing professional growth. Focusing on their classroom practice, teachers experiment with alternative approaches to engage students in mathematical ideas, possible strategies for assessment, and different methods of organization. They analyze and adapt strategies that they try, examining how well they help students develop mathematical competence and confidence. They incorporate such strategies into an ever-growing and more complex repertoire. Beyond the classroom walls, teachers also evolve as participants in a wider educational community. They read, work with colleagues, take the initiative to press for improvement, and raise their voices to speak out on issues affecting their

Teaching mathematics is a challenging profession and, as such, demands ongoing professional growth.

students, their schools, and their communities. Teachers' professional growth, within and outside their classrooms, is a product of their reflectiveness and of their participation in educational opportunities that will enhance and extend their growth and development (Artzt and Armour-Thomas 2002).

Among those opportunities is practice-based professional development, in which teachers investigate and analyze real situations of everyday teaching practice as a curriculum for professional learning (Ball and Cohen 1999). Those situations are studied through such artifacts as videotapes of a classroom or samples of student work. Teachers, acting as a collegial group, then investigate questions raised by the collected materials, for example, questions about students' learning of mathematics. The distinguishing characteristic of this type of professional development is that instead of having participants learn theories and apply them to the practice of teaching, it engages them in close examination of practice, from which theories or general principles emerge (Smith 2001).

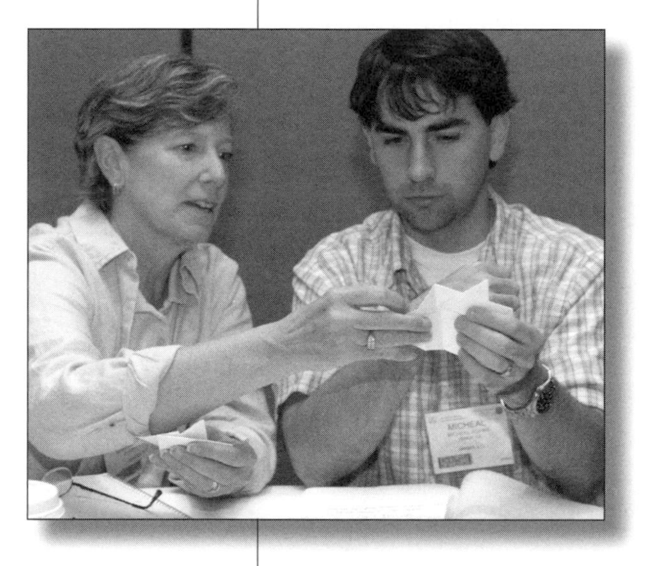

To implement a practice-based curriculum for professional learning, artifacts from classroom practice must be carefully selected with regard to their potential for inquiry. Also, professional tasks must be constructed around the artifacts, and the discourse must be facilitated in such a way that teachers are led to analyze, raise questions, experience disequilibrium, and progress in their practice.

Other professional learning strategies are organized and nurtured by collegial groups. One example is a group of colleagues who address issues of teaching and learning through reflection on professional readings. Members of book study groups read books or articles and then come together on a regular basis to analyze the material. Recognized as an important practice in teacher growth (Phi Delta Kappa Educational Foundation 2003; United Federation of Teachers Teacher Center 2002), the implementation of book study requires "a big shift from reflecting alone or with a partner to reflecting in a small group, such as a team or committee. While the potential impact of reflection increases, so does the personal risk" (York-Barr, Sommers, Ghere, and Montie 2001, p. 14).

Another example is a video or audio study group, which moves from the printed page to visual and audio input. In such groups, teachers reflect on and analyze teaching practice and student learning as they watch and listen to a classroom lesson as it unfolds in real time. Although commercially prepared packages serve such groups well by presenting carefully selected passages from a lesson, often with an accompanying guide to reflection, another way to use video offers other opportunities for growth. In some groups, teachers videotape their colleagues' or their own lessons, then gather to share, reflect on, and learn from those records of practice.

> [This type of professional activity] moves educators away from a view of teaching as a solitary activity, owned personally by each teacher. It moves them toward a view of teaching as a professional activity open to collective observations, study, and improvement. It invites ordinary teachers to recognize and accept the responsibility for improving not only their own practice, but the shared practice of the profession. For this new path to be traveled, however, teachers will need to open their classroom doors and, [instead of] evaluating each other, begin studying their practices as a professional responsibility common to all.
>
> (Hiebert, Gallimore, and Stigler 2003, p. 42)

Teachers will need to open their classroom doors and, [instead of] evaluating each other, begin studying their practices as a professional responsibility common to all.

Yet other collegial groups are committed to looking together at student work. Those groups develop structures and guidelines for analyzing small samples of student work, followed by reflection on significant questions about issues of teaching and learning. Examining student work focuses the conversation on student thinking and understanding. Teachers can hypothesize about student misunderstandings or communication concerns and can brainstorm with one another about ways to address those challenges.

In the collegial endeavor of Lesson Study, teachers design and test a lesson, observe the lesson being taught, then critically review it. Through a structured process, teachers discuss the lesson, not the teacher's performance, as they try to understand how students learned from the lesson. The one lesson, or series of lessons on the same specific topic, may be revised and retested throughout the school year. Although the goal may appear to be the creation of a highly polished lesson, as a total experience, Lesson Study is a strategy for professional development, and therein lies its value (Stigler and Hiebert 1999; Fernandez and Yoshida 2004).

Although all the foregoing strategies can be described as occurring in professional learning communities, the latter term is often used to denote a community of adults connected with a school's operation, including staff, parents, and students. Those communities, committed to school improvement as their overall goal, develop a shared vision and common values. A teacher's individual professional learning is defined within, and supported by, the group's mission, which includes a continuous commitment to improvement, as stated by DuFour and Eaker (1998):

> A commitment to continuous improvement is evident in an environment in which innovation and experimentation are viewed not as tasks to accomplish or projects to complete, but as ways of conducting day-to-day business, *forever*. Members of a professional learning community recognize and celebrate the fact that mission and vision are ideals that will never be fully realized, but must always be worked toward. (p. 28)

Other approaches to professional development, such as university courses, workshops, or conferences, depend on external qualified facilitators rather than collegial groups for their implementation. Learning experiences that immerse teachers in mathematical problem solving constitute one such approach. The teachers involved in such experiences focus on mathematical content, in depth, from a problem-solving perspective. Instructors model the facilitation of sharing a variety of solution strategies, with the intent that teachers will draw on those learning experiences in their own teaching practice. As teachers experience "teaching as less a matter of knowledge transfer and more an activity in which knowledge is generated through making sense of or understanding content, they begin to see their own role as teacher changing from a direct conveyor of knowledge to a guide helping students develop their own meaning from experiences" (Loucks-Horsley, Love, Stiles, Mundry, and Hewson 2003, pp. 195–96).

Space prohibits a full listing of all strategies for professional learning. However, teachers can take an active role in their professional development through such activities as—

- forming special-interest groups within their schools to investigate ways that technology might better enhance their teaching;
- participating in summer programs to explore such mathematics topics as statistics or algebra across grade levels;
- meeting with teachers from neighboring school districts to explore how they can jointly offer advanced mathematics courses for their students via telecommunications;
- working on curriculum renewal with other mathematics faculty to change the nature and kinds of courses that are being offered and to align their program with *Principles and Standards* (NCTM 2000);
- sharing strategies for implementing a newly adopted mathematics program; and
- joining local mathematics associations, attending meetings, making presentations, and assuming leadership roles.

As they find or create learning opportunities to enhance their practice, teachers as professionals take responsibility for their own growth and development.

Whether such opportunities are available to teachers depends to a significant extent on the school's support system, but it also depends on teachers themselves. As they find or create learning opportunities to enhance their practice, teachers as professionals take responsibility for their own growth and development. School systems must also recognize the importance of continued professional growth and support it by allotting incentives and, above all, the time needed to implement it.

Teachers as Agents of Change

Among the many voices expressing concerns and opinions about the teaching and learning of mathematics, one voice should receive particular respect and attention: the voice of the informed mathematics educator. With this voice, mathematics educators are a resource for parents, administrators, colleagues, and other concerned citizens regarding issues in mathematics education. Mathematics educators should be informed about multiple perspectives on issues of curriculum and pedagogy: rationales for various programs and approaches together with potential benefits and limitations, decisions made by one's own school or district, as well as other resources for more information.

What is a profession? Shulman (1998) suggests that all professions are characterized by a number of attributes including—

- an obligation to the service of others;
- a commitment to generate and disseminate a scholarly body of knowledge;
- a need for continued growth through the intersection of theory and practice; and
- a connection with a collegial community that values and encourages quality in performance of tasks associated with the profession.

Professionalism among teachers is built through a support system that links them with colleagues inside and outside the schools. Teachers should be able to turn to colleagues for information concerning any aspect of mathematics education to expand their views of mathematics, their resources for teaching, and their repertoire of teaching and learning skills. Such interchange fosters intellectual refreshment and places teachers in the role of partners in the process of education. It also offers opportunities for heightened awareness of teachers' responsibility for fostering their own professionalism by building collegial networks, reading professional literature, becoming involved with professional organizations, and initiating contact with teacher educators at local colleges and universities.

Professionalism among teachers is built through a support system that links them with colleagues inside and outside the schools.

Teachers who are proactively engaged in improving mathematics education demonstrate the notion of professionalism in many ways. Essential to their success in that proactive role is that teachers view themselves as agents of change, responsible for improving mathematics education at many different levels: the classroom, the school, the district, the region, and the nation.

Standard 5: Participation in Career-Long Professional Growth

Vignette

In this vignette, teachers consider two ways of introducing a problem and guiding students through its solution. They relate their own experiences in solving the problem to the experiences of students learning in different ways. By having this discussion in a community of practitioners, the teachers can reflect together on the pros and cons of different instructional methods.

5.1—*Rethinking Teaching Strategies: The Condominium Problem*

Ms. Costa, the director of a local state-supported teachers' project, wants to encourage participating teachers to examine their assumptions about the nature of mathematics and how it should be taught. In particular, she wants them to appreciate the difference between understanding how to solve a problem and merely being able to apply an algorithm. At a meeting she poses a problem not typical of those that appear in most textbooks:

She wants them to appreciate the difference between understanding how to solve a problem and merely being able to apply an algorithm.

> In a certain adult condominium community, two-thirds of the male residents are married to three-fifths of the female residents. What part of the community is married?

Bob: There isn't enough information.

Tony: Are the people in the community only married to others in the community?

Connie: We don't know how many people there are. Are there the same number of men and women?

Ms. Costa: You raise some good issues and questions. What are some conditions that aren't explicitly stated but that we would assume in order to make sense of the problem?

Tom: Men and women in the community who are married are only married to others in the community.

Connie: The number of men is the same as the number of women.

Susan: I don't think that's true. The number of married men should be the same as the number of married women, but the totals could be different.

The teachers continue to list several assumptions, selecting those with which they agree and recognizing others as misleading.

The teachers work on the problem in small groups while Ms. Costa walks around the room. She notices that some people are still clarifying issues raised in the initial discussion while others have agreed on conditions and are trying to determine what part of the community is married.

A variety of approaches emerge from the small-group discussions. The participants later share those approaches with the class. Ms. Costa encourages to them to generalize by asking, "How are the solution strategies alike, and how are they different?" "Do they all produce the same result?" "What if the fractions in the original problem are changed?"

Ms. Costa: Now suppose this sort of problem is included as part of the school mathematics curriculum. I'd like to demonstrate a possible way to present the problem and solution to students.

Ms. Costa then walks the teachers through a rule-based approach to solving the problem. She develops a three-step procedure and explains exactly what calculations should be done at each step.

> Step 1. Find the L.C.N. (least common numerator). [The L.C.N. for 2 and 3 is 6.]

Step 2. Change fractions to equivalent fractions with the same L.C.N. [2/3 = 6/9 and 3/5 = 6/10.]

The final step is to use the "add" method of combining fractions. In this procedure we combine two fractions by adding the numerators to get the new numerator and the denominators to get the new denominator.

Step 3. Add the two fractions. [6/9 + 6/10 = 12/19.]

Ms. Costa: These three simple steps will allow you to solve all "marriage type" problems.

Ms. Costa visually surveys the participants to observe their reactions. Some participants seem surprised, others are confused, and still others are nodding in agreement with the answer and procedure.

Ms. Costa: What do you think? It didn't take very long and could be accomplished quite easily in the school curriculum. Of course, you would have to spend time memorizing the three steps and practicing them. How does the method compare with the methods you used earlier to solve the problem?

Joe: It's quick, but it doesn't explain what you're doing.

Marissa: It was really the same as my group did before, only more formal. We found the same numerator. Then the total number of each is the denominators. But we just counted everything to get the totals.

Ms. Costa: Your group tried to make sense of the three steps. Do you think you could have done that without having had time to work on the problem on your own and compare the various methods we came up with before?

Tony: I don't think so, but it sure seems easier for the kids to follow the three steps.

Ms. Costa: Well, I could have started today's lesson by presenting the problem and giving you the three steps to solve it. We could have spent time practicing the steps and doing similar problems. You could have been quite successful in solving problems of that type without understanding much about why you were doing the things you did. In the short run you would appear to be successful, but in the long run where would you be?

Do you think you would remember this formula several years later if you had learned it by rote and you encountered a similar problem? Might you also have been taught other, similar formulas by rote that you might confuse with this one?

Ms. Costa encourages the participants to return to their groups to discuss the following questions: Which approach to teaching the formula will lead to true, lasting understanding by their students? How does having students memorize formulas by rote in the interest of "covering" the curriculum do a disservice to students? In what ways can classroom teachers give their students rich, problem-solving-based learning experiences, yet accomplish their school's or district's curricular objectives.

Vignette

In the next vignette, a teacher questions some aspects of her mathematics instruction and takes steps to find help in improving her practice. With leadership from a supportive principal and the encouragement of another teacher, the teacher becomes better able to analyze and reflect on events in her own classroom. The collaborative process eventually leads to the teacher's involvement in a broader professional development project.

5.2—*Redesigning Instruction: Collegial Support*

Stacy Washington, a fourth- and fifth-grade teacher for three years, has grown increasingly dissatisfied with her mathematics teaching. She feels that she has not made the subject come alive for her students. Stacy believes that her students should understand the mathematics they are doing, learn to reason mathematically, and use their knowledge to solve problems. However, they must spend so much time working alone to learn the mathematical rules and procedures that no time seems to be available for lively group discussions.

Stacy has tried to use concrete materials as tools in her teaching but has felt as if her students have not reaped the benefits she expected from working with manipulatives. She did use fraction bars when teaching addition and subtraction of fractions, but they just served as tools to help get the right answer. Furthermore, that particular representation completely broke down for her when she introduced multiplication of fractions.

Determined to get help, Stacy expresses her concerns to her principal, LaTasha Enary-Fayse, who suggests that she talk with other teachers who have been working on similar issues. After making a few telephone calls, LaTasha gives Stacy the name of a teacher in a neighboring district who has made changes in his mathematics teaching over the past few years. She agrees to arrange for a substitute teacher once Stacy has scheduled a visit.

Stacy's visit with Jon Nickerson proves to be fruitful. She observe that his mixed-ability seventh-grade students are not only engaged in sophisticated and serious dialogue about mathematics but are clearly developing basic skills through discovering patterns, articulating them to the class, and determining whether they can be generalized.

While comparing the graphs of four different linear equations on the computer, one student noticed that the two lines with "the same number in front of the x were parallel."

The students have not been taught the definition of slope, but they are discovering patterns that relate to the slope of the lines. Jon asks such questions as "Will this always be true? How do you know? Do you agree? Why?"

Stacy tries to imagine her students in a similar discussion. She realizes that she knows little about what her students are thinking about the mathematics they are learning.

Jon's teaching is not composed of techniques that Stacy can assimilate through a single observation. Afterward, she has the opportunity to talk to Jon and finds that he understands her concerns. The follow-up discussion of issues and concerns gives Stacy insights into the process by which Jon reflects on and modifies his teaching.

As a mathematics major in college, Jon discovered firsthand that learning mathematics involves more than memorizing rules and practicing them. However, the translation from his own experience as a learner to his strategies as a teacher was not direct. Jon had to try a variety of approaches to learn how to foster the sort of inquiry and discourse that Stacy had observed in his classroom. "In fact, I am always learning more and trying out new ideas," he explains.

Jon shows Stacy his copy of the NCTM *Principles and Standards for School Mathematics* (2000) and encourages her to join NCTM to take advantage of their regular publications. As they look through the document together, she is struck by the emphasis on understanding that goes beyond getting right answers. They discuss several strategies that Stacy might use in her class to start things off.

Stacy leaves convinced that her students should explain their answers and that they should learn to listen to, and make sense of, other students' solutions. She and Jon agree to meet regularly to share their progress and struggles. They plan to invite their colleagues to join them in their discussions. Jon also offers to observe Stacy's class when she feels ready.

Throughout the rest of the year, Stacy's progress is gradual and steady. Eventually, Stacy is comfortable having Jon and LaTasha visit her classroom. LaTasha also asks whether she may arrange for other teachers to observe some of Stacy's lessons.

Toward the end of the year, LaTasha asks Stacy if she would be willing to work with other mathematics teachers in the district to make curriculum decisions and plan professional development opportunities. Stacy happily agrees and is looking forward to learning from the other teachers. She is surprised to realize that she had never thought of herself as a mathematics teacher before, even though she teaches it every day.

Vignette

In the next vignette, teachers and university faculty work together to learn more about potential uses of interactive geometry software. In particular, they discuss software features that represent algebraic functions dynamically. Wary of technology's potential pitfalls, they seek out activities that exploit the power of technology to enhance students' understanding of concepts and skills that they have already determined are important.

5.3—*Expanding Perspectives on Technology*

A group of high school mathematics teachers has been meeting twice a month at their school for a seminar with mathematicians and mathematics educators from a nearby university. The teachers have had computers in their schools for quite a while but have not felt they were taking full advantage of the resources available to them. They want to learn more about using interactive geometry software in their classes.

They want to be sure that they use the power of technology to enhance mathematical understanding, not simply for the sake of using technology.

The teachers are eager to find appropriate tasks for the interactive geometry environment that take advantage of the technology in a way that enhances understanding. They have heard that interactive geometry software can be used in teaching algebra and other topics, but they are not sure where to begin. They are also sensitive to criticisms of technology and want to be sure that they use the power of technology to enhance mathematical understanding, not simply for the sake of using technology.

Brad: I've used [interactive] geometry software for years to help students make conjectures about geometric relationships. Sometimes I use activities that I find in books. Other times, I modify activities or make up my own. I always try to have students write some kind of proof or justification for their conjectures and share those during the next class period. It helps hold them accountable for what we do in the lab and reminds them that mathematics isn't just about making conjectures.

Andrea: I saw another way to use [intertactive] geometry software at a professional conference. The presenter showed how to manipulate a function by varying the parameters. But unlike a graphing calculator, the function did not have to be re-entered to see the effect of a change in the parameters. There were sliders, or line segments, that indicated the values of the parameters. The distance between a slider's origin point and its endpoint corresponded to the magnitude, or absolute value, of the parameter. Moving the endpoint of the slider to the right of its origin point made the value of the parameter positive. Dragging the endpoint of the slider to the left of its origin point resulted in a negative value of the parameter. The dynamic aspect of this was that as you changed the length of the slider, the graph of the function changed dynamically as well. It was really impressive.

Jong-Ho: I have also seen an interesting way to represent functions that was called a *dynagraph* In some ways, it is similar to what Andrea was describing. There were two axes representing the values of the independent variable and the dependent variable, respectively. But unlike Cartesian coordinates, or *x-y* axes, which are perpendicular to one another, these axes were parallel number lines. There was an indicator on the independent-variable axis that was connected to a different indicator on the dependent-variable axis. As you slide the indicator across the independent-variable axis, the indicator on the dependent-variable axis moves according to a hidden function that is governing its behavior. So, in some sense, the independent-variable axis is like an input slider, like Andrea was describing. But the output is dynamically represented along a one-dimensional number line. You can really see how a linear function differs from a quadratic function. And you really have to think hard to figure out what kind of function you are looking at.

Dr. Papa: I've used some similar aspects of [interactive] geometry software to help students see the relationships between functions and their derivatives when I've taught calculus at the university. The visualization aspect is very powerful and helps set up the conceptual understanding and connections with the symbolic representations.

Brad: These applications seem very interesting and may have lots of potential to help students make connections among representations. When I use these kinds of activities, however, I want to be sure that I have a clear purpose in mind, and that I can assess the value of the tasks in terms of my students' understanding of the concepts I want them to learn.

Andrea: I totally agree. I think we should make a list of concepts and skills that we plan to teach related to functions and see if any of these tools can be used to help students develop in thse areas.

Jong-Ho: Wait a minute. I'm not sure I'm prepared to do that with as little as I know about sliders and dynagraphs.

Dr. Feeney suggests that at their next meeting the group spend some time in the lab exploring some of the features of the software that relate to functions. She asks each of the group members to look ahead in their courses and bring back a list of objectives related to functions. She also promises to work with Dr. Papa and others to compile a list of resources and lesson plans that they can work through in the lab and discuss with respect to the objectives that the teachers bring with them.

Closing Thoughts
Standard 5: Participation in Career-Long Professional Growth

This standard describes the role that teachers should play in their own ongoing professional development. In particular, teachers should continually observe and analyze a variety of teaching and assessment practices as well as examine the assumptions about teaching and learning underlying each of them. They should participate in small and large groups of other professionals who are reading books, viewing videotapes, or exploring technologies in search of appropriate and effective teaching methods. As informed educators, teachers should also take on the role of advocates for the improvement of mathematics teaching and learning, and they should serve as resources for others who are interested in issues related to mathematics instruction and assessment.

Summary of the Education and Continued Professional Growth Standards

These Standards describe teacher development as an ongoing process, delineating the learning experiences needed at various stages and the opportunities that should be available to stimulate professional growth. Also outlined are the specialized types of knowledge—knowledge of mathematics, knowledge of learners, and knowledge of mathematics pedagogy—required by teachers to effectively help students learn mathematics. Finally, these Standards address the responsibility of mathematics teachers to take an active role in their own professional development. Throughout the education and continued professional growth Standards, we have emphasized the importance of collegial support in the processes of learning, reflecting, and working to improve instruction and assessment.

Chapter 4

Working Together to Achieve the Vision

Mathematics Teaching Today: Improving Practice, Improving Student Learning presents a vision that calls for a teacher who is educated, supervised, mentored, and supported in ways that require collegiality and professionalism from fellow teachers as well as from many others who influence the work of teachers. To create teaching and learning environments that support mathematical inquiry and decision making through mathematical problem solving, communicating, reasoning, using multiple representations, and connecting ideas, teachers must have access throughout their professional lives to educational opportunities that focus on developing a deep knowledge of subject matter, pedagogy, and students. Many external forces and decisions affect mathematics teaching and school mathematics programs. Various constituencies have responsibility for supporting mathematics teachers and teaching and for building successful mathematics programs. Such support needs to be multifaceted, systemic, and reliable.

Many existing support systems for mathematics teachers are as inadequate for teaching in today's society as the shopkeeper arithmetic curriculum is for educating our children to live and work in the twenty-first century. The kind and level of mathematics education required for today's students to prosper in a dramatically changed economy and in a scientifically and technologically advanced society places great responsibility on the shoulders of teachers of mathematics. At the same time our society is undergoing other dramatic changes that make teaching even more challenging. We are growing more diverse along many dimensions: ethnically, culturally, linguistically, in family patterns, in the integration of persons with disabilities into mainstream institutions, and in numerous other ways (U.S. Census Bureau 2001; Farberman 2006).

Responsibilities for the Support and Development of Teachers and Teaching

Teachers can and do implement successful mathematics programs with little help or encouragement. However, such practice should not be expected to flourish without adequate support. The changes called for by NCTM, including recommendations made in *Principles and Standards for School Mathematics (Principles and Standards)* (2000) and other *Standards* documents, need the support of many segments of society. This section summarizes the responsibilities of six major constituencies for the support and development of teachers and teaching:

Many external forces and decisions affect mathematics teaching and school mathematics programs.

1. Responsibilities of Policymakers in Local, State, Provincial and National Governments
2. Responsibilities of Business and Industry Leaders
3. Responsibilities of Schools and School Systems
4. Responsibilities of Colleges and Universities
5. Responsibilities of Professional Organizations
6. Responsibilities of Families and the Community

Each of these individuals and groups has responsibilities to help support and shape the environment in which teachers teach and students learn mathematics. In addition to identifying specific responsibilities, this section also discuss the process of implementing reform in mathematics education and the roles of various groups in that process.

Responsibilities of Policymakers in Local, State, Provincial, and National Governments

Policymakers in government should take an active role in supporting mathematics education by accepting responsibility for—

- supporting decisions made by the mathematics education professional community that set directions for mathematics curriculum, instruction, evaluation, and school practice; and
- providing resources and funding for the implementation of high-quality school mathematics programs that reach all students, as envisioned in *Principles and Standards* and other NCTM *Standards* documents.

Elaboration

Policymakers must confer with and support teachers and other mathematics education professionals on issues that affect what and how a teacher of mathematics can teach.

Policy decisions are made at many levels that affect the status of teachers and teaching and the environment in which teachers teach mathematics. High school graduation requirements; state, provincial, or federally mandated pupil and teacher testing; state or provincial department of education certification requirements; textbooks published and adopted; standardized tests published and adopted; local scheduling of classes; teaching assignments; allocation of resources; policies that affect professional development, such as attendance at professional meetings; teacher evaluation procedures—these are but a few of the myriad decisions that either lend support to, or substantially constrain the improvement of, mathematics teaching and learning. All too often such policy decisions are made without consultation with teachers and other mathematics education professionals; yet wise policy decisions that affect the mathematics education of all students must reflect the education, experience, and expertise of the mathematics education community. Teachers at all levels are held accountable for the mathematical growth of students. Therefore, to be effective, policymakers in all arenas—state, provincial, local, national—must confer with and support teachers and other mathematics education professionals on issues that affect what and how a teacher of mathematics can teach.

Providing all students the opportunity to solve problems, reason, make connections, communicate, and use a variety of representations requires a variety of resources. According to *Mathematics Benchmarking Report, TIMSS 1999—Eighth Grade* (Mullis et al. 2001), students in well-resourced schools have higher mathematics achievement. Policymakers in government must understand the need for materials and tools for learning and doing mathematics and must ensure that adequate and appropriate resources are available for teachers and students. Such resources include physical space, books, supplementary text materials, technology, software, manipulative materials, supplies, library materials, and audiovisual resources. In addition, classroom materials should be chosen by teachers according to the needs of their students and in alignment with the NCTM *Standards* documents.

Students in well-resourced schools have higher mathematics achievement.

Policymakers must understand the mathematical needs of workers and citizens of the future and must join to help make such a mathematics education a reality for all students. If teachers are to be able to realize such a goal in the classroom, they must have the financial and other support to continue to learn. They must have the time to reflect on their teaching and on students' understandings so that they can reach all students (Artzt and Armour-Thomas 2002). In fact, teacher improvement depends on adequate time and resources for continual learning and sharing of professional knowledge (National Commission on Mathematics and Science Teaching for the 21st Century 2000; Stigler and Hiebert 1999) and will require dramatic restructuring of teaching assignments and provision of ongoing professional development opportunities for teachers of mathematics. Such restructuring can be accomplished only with the support and involvement of policymakers in government. Schools, teachers, and students—in fact, our society—need and deserve such help.

Teacher improvement depends on adequate time and resources for continual learning and will require dramatic restructuring of teaching assignments and provision of ongoing professional development opportunities.

Responsibilities of Business and Industry Leaders

Business and industry leaders should take an active role in supporting mathematics education by accepting responsibility for—

- providing assistance in, developing, and implementing high-quality school mathematics programs that reach all students, as envisioned in *Principles and Standards* and this volume; and
- participating in partnerships at the national, state or provincial, and local levels to improve the teaching and learning of mathematics.

Elaboration

Businesses that have a particular responsibility in the reform effort are publishers of textbooks and standardized tests. We know that textbooks and tests have a profound influence on what is taught. Therefore, authors and publishers have both an

Textbooks and tests have a profound influence on what is taught.

opportunity and a responsibility to help improve mathematics instruction. Publishers should seek advice from teachers and other mathematics educators in making decisions that affect what is developed and published in mathematics. Teachers have a responsibility to insist on materials that are the best suited for children in learning mathematics and on testing that is aligned with the goals of the mathematics program.

Business and industry leaders should realize that they stand to gain substantially from excellent school mathematics programs. Large amounts of money are spent each year to train workers in the mathematics needed in the technological workplace (National Commission on Mathematics and Science Teaching for the 21st Century 2000). Having a work force that is better educated in mathematics in prekindergarten–grade16 programs would change the nature of on-the-job training and increase economic competitiveness. Therefore, business and industry must join with communities and schools in improving mathematics instruction. The kinds of help needed are both financial and human. Supporting schools with resource materials, especially technological resources, is one way that businesses can help prepare students for work in a technology-driven society. From a human resource perspective, arranging for scientists, mathematicians, and other business professionals to spend time in schools can be very beneficial in stimulating students to study more mathematics, especially when the visiting business and industry personnel are culturally diverse. Interacting with members of traditionally underrepresented groups, such as African Americans, Hispanics/Latinos, First Nations People, Native Americans, Alaskan Natives, Pacific Islanders, Asian Americans, females, or disabled professionals, helps both students and teachers set higher expectations for all students. Participation in programs that grant teachers internships in business and industry can stimulate teachers' professional growth and help them bring the reality of the uses of mathematics in the workplace to their students. Such programs also send signals that mathematics teaching and teachers are valued in our society, thus making mathematics teaching a more attractive career option for talented young people.

Responsibilities of Schools and School Systems

School administrators and school board members should take an active role in supporting teachers of mathematics by accepting responsibility for—

- understanding the goals for the mathematics education of all students set forth in *Principles and Standards* and the needs of teachers of mathematics in realizing those goals in their classrooms;
- recruiting qualified teachers of mathematics, with particular focus on the need for a diverse teaching staff;
- providing a mentoring and support system for beginning and experienced teachers of mathematics to ensure that they grow professionally and are encouraged to remain in teaching;
- making teaching assignments on the basis of qualifications of teachers;

- involving teachers centrally in designing and evaluating programs for professional development specific to mathematics;
- supporting teachers in self-evaluation and in analyzing, evaluating, and improving their teaching with colleagues and supervisors;
- providing adequate resources, equipment, software, instructional materials, time, and funding to support the teaching and learning of mathematics as envisioned in this document;
- allowing teachers input in selecting curriculum materials, software, technology, and other resources that focus on learning with understanding, and providing the time, professional training, and financial resources needed to help teachers implement such materials in ways that maximize their effectiveness.
- establishing outreach activities with parents, guardians, leaders in business and industry, and others in the community to build support for high-quality mathematics programs; and
- promoting excellence in teaching mathematics by establishing an adequate reward system, including salary, promotion, and working conditions.

Elaboration

Principles and Standards, other NCTM *Standards* documents, and this volume lay out a vision for school mathematics and instruction that should become the framework for planning school mathematics programs. Principals, superintendents, and other administrators must understand these documents so that they can represent the mathematics program to the community, especially to families, in ways that help generate the support that teachers need to implement a high-quality mathematics program that meets the needs of every student. School administrators must understand that the mathematics education called for in these Standards requires time for mathematics; time for teachers to plan, reflect, and collaborate for the purpose of improving instruction; time for engagement in professional growth activities; and time to interact with the community. The payoff for such commitment is the realization of a mathematics program that offers all students the opportunity to solve problems, reason, make connections, communicate, and use a variety of representations in mathematics.

As our society grows more diverse ethnically, culturally, and linguistically, and in the integration of people with disabilities into mainstream institutions, the teaching force must also change to reflect this diversity. For example, a school environment that includes teachers of mathematics who are African Americans, Hispanics/Latinos, First Nations People, Native Americans, Alaskan Natives, Pacific Islanders, Asian Americans, or females, or who have disabilities helps children of those cultural, ethnic, gender, or ability groups see themselves as capable members of society who can do significant mathematics. To achieve such diversity among teachers of mathematics requires aggressive, systematic, and long-term support for the young people who are recruited into the mathematics teaching profession.

As our society grows more diverse, the teaching force must also change to reflect this diversity.

Teaching mathematics, as with any other subject, requires knowledge and experience

Teaching mathematics, as with any other subject, requires knowledge and experience that is specific to that discipline.

that is specific to that discipline. Although discussions of knowledge, tasks, discourse, environment, and analysis are appropriate in relation to any school subject, a professional teacher's knowledge of how students learn, of the subject-matter content, and of the pedagogy are specific to that particular discipline. An excellent social studies teacher should not be assigned to teach a portion of the mathematics curriculum under the assumption that his or her teaching knowledge and skills will transfer. Teaching assignments must be made on the basis of the qualifications of teachers.

Schools have major responsibilities for fostering the professional growth of teachers of mathematics. Their work in supporting appropriate professional development programs, promoting collegiality, and recognizing the role of teachers as responsible professionals does much to enhance high-quality programs and instruction in mathematics education. Such professional growth experiences should begin as soon as teachers enter the profession. Once in the profession, beginning teachers need supportive guidance to help them develop their skills and habits of mind as teachers of mathematics. Such programs for beginning teachers are essential to retain talented young people in the very demanding, as well as rewarding, profession of teaching mathematics.

The unique challenges facing teachers of mathematics are best addressed when teachers are given opportunities to engage in ongoing, subject-specific professional development programs. Their concerns and interests are met when teachers play a major role in identifying and assessing their own needs. Planning and developing continuing education programs should take place within individual schools and school districts and should highlight teacher involvement in both program development and program implementation (Loucks-Horsley, Love, Stiles, Mundry, and Hewson 2003).

Professional programs require the commitment of resources, equipment, time, and funding. For example, the ready availability of technology in the workplace and the implications of that resource for use in mathematics teaching and learning underscore the crucial need for including technological equipment as instructional and learning tools. Schools should allocate a fixed percent of their budgets for faculty development activities.

Principals should allocate time for teachers to build collegial links with other faculty.

As a part of professional development, principals should allocate time for teachers to build collegial links with other faculty. Teachers need opportunities to share ideas, plan interdisciplinary lessons, and explore instructional strategies. Returning to the university to pursue graduate programs in mathematics education allows teachers to deepen their knowledge and skills in mathematics and mathematical pedagogy. Schools that are supportive of such activities for teachers benefit from a teaching staff that continues to learn and to improve the mathematics program and students' successes in mathematics.

The goal of more and better mathematics for all students cannot be reached with chalkboards and worksheets, inadequate class time, and six classes a day per teacher. Mathematics teachers need appropriate resources. Calculators, computers, and manipulatives are as integral to learning and doing mathematics as chemicals are to a chemistry laboratory. If mathematics teachers are to meet the needs of the increasingly diverse student populations in their classrooms, then released time to plan, to study, to reflect, to develop curriculum, and to confer with colleagues is essential.

The goal of more and better mathematics for all students cannot be reached with chalkboards and worksheets, inadequate class time, and six classes a day per teacher.

High-quality mathematics education must be a concept to which we commit for all students, not just for a privileged few. This goal requires a commitment from the school and community for adequate funding to support the teaching and learning of mathematics. However, for members of the community to be supportive of the mathematics program, they must know the goals of the program and must understand the kind of support needed by teachers to carry out the program. In this endeavor, school administrators, especially principals, play crucial roles. Principals who take the time to work with their teachers in developing a coherent, powerful mathematics program and, further, take the responsibility to be the advocate for mathematics teachers in the community, can make it possible for teachers to teach and for students to learn as envisioned in *Principles and Standards,* other NCTM *Standards* documents, and this volume. For teachers to move toward the vision of teaching in these Standards, school administrators must establish a reward system, including salary and promotion, that supports and encourages teachers as they grow professionally. One productive and rewarding way to support teachers in making improvements in the mathematics program is to fund extended-year contracts for summer pay to develop special, teacher-proposed projects.

School administrators must establish a reward system, including salary and promotion, that supports and encourages teachers as they grow professionally.

Responsibilities of Colleges and Universities

College and university administrators need to take an active role in supporting mathematics faculty, mathematics education faculty, and teacher development programs by accepting responsibility for—

- establishing an adequate reward system, including salary, promotion and tenure;
- providing a workplace environment that supports teacher educators in all aspects of their professional practice;
- facilitating faculty collaboration that focuses on the ongoing improvement of teacher development programs; and
- ensuring that adequately qualified individuals are assigned to teach methods and content courses to teachers or teacher candidates

so that faculty can and are encouraged to—

- spend time in schools working with teachers and students;
- collaborate with school teachers and other school-level personnel in the design and implementation of school-based preservice and continuing education programs;
- engage in inquiry-oriented analyses of their own practices to stimulate professional growth and program enhancement;

- offer appropriate graduate courses and programs for experienced teachers of mathematics;
- provide leadership in conducting and interpreting mathematics education research, particularly school-based research, as well as collaborating with school-level personnel on their own action research;
- cooperate with precollege educators to articulate prekindergarten–grade 16 mathematics programs; and
- make concerted efforts to recruit and retain high-quality, diverse teacher candidates.

Elaboration

The work of colleges and universities is fundamental to successful mathematics teaching and the education of qualified mathematics teachers. Colleges and universities have major responsibilities not only for the preservice and continuing education of teachers but also for relevant programs that are responsive to today's and tomorrow's educational needs. Faculty should collaborate with other practicing professionals to design preservice and continuing education programs that reflect NCTM *Standards*–based models of teaching and learning.

Just as school-level teachers should engage in inquiry-based reflection on their own practices, so should college and university faculty examine their own teaching and its effect on teachers' developing understandings and perspectives. This type of analysis is essential because of the mentoring role that college and university faculty have as they model the teaching of mathematics. Mathematics educators, mathematicians, and others should collaborate in such efforts to improve their teaching by sharing experiences, perspectives, and expertise in a spirit of collegiality. To support faculty efforts to provide the best possible experiences for teachers, colleges and universities should ensure that only highly qualified individuals are assigned to teach both content and methods courses.

Flexible and alternative methods for continuing education and self-improvement must be instituted to support ongoing learning of mathematics and mathematics education. Graduate programs offered in the late afternoon and during the summer make continued study more accessible for teachers. Such learning opportunities need to be particularly attentive to the special needs of adult learners—the practicing mathematics professionals.

Colleges and universities must also work with schools to initiate recruitment and retention efforts to attract high-quality candidates to the field of mathematics teaching. Efforts should be established to attract, support, and retain persons from traditionally underrepresented groups in teaching.

Mathematics educators should participate in school mathematics programs in ways that keep them current with respect to experience of, and knowledge about, the realities of classroom instruction. The supervision of student teachers requires working closely with cooperating teachers and preservice teachers for significant time periods. During those times, mathematics educators not only observe the student teacher but gain an understanding of all aspects of the school environment. On another level, mathematics education faculty can work with classroom teachers in their classes so as to gain firsthand experience with teaching in today's classrooms and to have an environment in which to implement current, research-supported teaching practices and learning principles.

Mathematics educators can also work in collegial partnerships with school-based personnel, such as in professional development schools, or by supporting teachers' action research. Such work is recognized as part of the full teaching and research responsibilities required by the college or university and is usually welcomed and facilitated by a local school district. Collegiate faculty can use such opportunities to offer valuable, supervised classroom teaching experiences to the preservice teachers with whom they are working. Those preservice teachers, in turn, learn by having opportunities to try a variety of pedagogical approaches in a supported environment. Collegiate faculty both influence and are influenced by their ongoing interaction with prekindergarten–grade 12 students, with experienced teachers, and with preservice teachers at their college or university. This involvement with schools is necessary to help articulate mathematics programs for prekindergarten–grade16.

Colleges and universities have a major responsibility to work with and in schools to develop new knowledge to shape practice. Educational professionals can employ basic and applied research in the teaching and learning of mathematics to gain both theoretical and practical knowledge to guide mathematics teaching. Teachers must be recognized as and encouraged to be partners with college and university faculty in planning, conducting, and interpreting research that affects mathematics teaching and learning.

Collegiate faculty should be actively involved in the mathematics and mathematics education communities through participation in a variety of professional organizations. Such participation enables them to share their expertise by making professional contributions to publications and conferences.

For collegiate faculty to meet the responsibilities outlined in this section, college and university administrators must establish a reward system, including salary, promotion, and tenure, that supports and encourages faculty to engage with schools in such work. This support calls for a change in the current culture of colleges and universities. Thoughtful analytic work in schools must be recognized as a scholarly activity that is demanding both physically and intellectually and that is crucially important to society. In many departments of mathematics, a large percent of the students who are mathematics majors are preservice teachers, yet in the current culture of those departments, the preservice teachers and the faculty who work with them are held in lower regard than their fellow students and colleagues. The mathematics education community must

College and university administrators must establish a reward system, including salary, promotion, and tenure, that supports and encourages faculty to engage with schools.

continue to work diligently to change those perceptions, as well as to reverse the negative attitudes toward mathematics held by many members of society.

Responsibilities of Professional Organizations

The leaders of professional organizations need to take an active role in supporting teachers of mathematics by accepting responsibility for—

- promoting and providing professional growth opportunities for those involved in mathematics education;
- focusing attention of the membership and the broader community on contemporary issues dealing with teaching and learning mathematics;
- promoting activities that recognize the achievements and contributions of exemplary mathematics teachers and programs; and
- initiating political efforts that effect positive change in mathematics education.

Elaboration

Professional organizations foster a strong sense of community through various strategies, such as written materials, videotapes, journals, and conferences. Such vehicles allow teachers of mathematics to link with other professionals through the growing use of a variety of electronic communication media.

Demands for improved mathematics education have been well documented in many reports. The mathematics education community should take the lead in the reform efforts, and toward that end, professional organizations are essential in helping mathematics educators be heard in the vast political community. This advocacy role necessitates strong, vital organizations that can inform mathematics educators of current issues, encourage attention to diverse points of view about important content and methodology, mobilize efforts to reach consensus on needed outcomes, present a strong and unified voice for dynamic and thoughtful change, and make this voice heard in the larger corporate, government, and policy-making sectors.

The work of local, state, provincial, regional, and national organizations supports teachers in finding avenues to be involved in decision making that affects mathematics education. The work of NCTM on the growing collection of *Standards* documents and resource materials, as well as efforts by members of the American Mathematical Society and the Mathematics Association of America on comprehensive reports, such as the Conference Board of the Mathematical Sciences' (CBMS) *The Mathematical Education of Teachers* (CBMS 2001), has involved thousands of mathematics educators in drafting, reviewing, and revising standards for practice. This activity is a prototype for arriving at professional consensus to give definitive direction to practitioners in important aspects of a discipline.

All organizations are challenged to assert their leadership roles in supporting and encouraging mathematics teachers to grow professionally and to achieve greater recognition as respected professionals. The increased status of mathematics educators will influence new candidates to choose teaching as a rewarding lifetime career. Above all, professional organizations must strive for society's recognition of teachers as the professionals they are.

All organizations are challenged to assert their leadership roles in supporting and encouraging mathematics teachers to grow professionally and to achieve greater recognition as respected professionals.

Responsibilities of Families and the Community

Families and community members should take an active role in supporting teachers of mathematics by accepting responsibility for—

- maintaining a positive attitude about the value of learning and understanding mathematics and for expecting that all children can learn mathematics;
- providing experiences for young children that build on their natural curiosity about the world and that support and encourage mathematical exploration;
- understanding the goals for the mathematics education of all students set forth in *Principles and Standards* (NCTM 2000) and the needs of mathematics teachers in realizing those goals in their classrooms; and
- supporting mathematics teachers and promoting mathematics teaching as a profession.

Elaboration

One of the curious aspects of our society is that it grants social acceptability to taking pride in not being good in mathematics. A phrase often uttered by many members of society, including some teachers, is "Oh, I was never any good at mathematics." In many other societies the prevailing assumptions are that all students can learn mathematics and that learning is a matter of effort. In our society, we are more likely to think that persons are either born with a mathematical mind or they are not. Our expectations have a great deal to do with how we respond to students and consequently to what students believe they can do. Families and community members, as well as teachers, counselors, school administrators, and students, need to have high expectations that every student can learn mathematics. Because motivation and positive attitudes toward mathematics are linked with achievement, positive messages and encouragement are important aspects of school and home environments (Organisation for Economic Cooperation and Development 2006). The NCTM *Standards* documents challenge us to create learning environments for students and for teachers in which the primary goal is building students' confidence in learning and doing mathematics.

Families and community members, as well as teachers, counselors, school administrators, and students, need to have high expectations that every student can learn mathematics.

Families should provide high quality, challenging, and accessible mathematics education for three- to six-year-old children because it is a vital foundation for future mathematics learning (National Council of Teachers of Mathematics and National Association for the Education of Young Children 2002). Throughout the early years of life, children have a natural tendency to notice and explore mathematical aspects of their world. Children compare quantities, such as bigger or smaller, more or less. They observe and create patterns. Their curiosity about the three-dimensional world we live in also leads children to become familiar with solid objects and explore issues of balance and space as they play with blocks of various sizes and shapes. As children count and share toys or food, they lay the foundations for doing computations later on. Children use mathematics to make sense of their world outside of school and, as a result, develop a solid foundation for more formal mathematics that they will learn in school.

One of the biggest challenges to reform is the tendency for a system to maintain the status quo. The NCTM *Standards* documents and companion resources offer a vision of mathematics education that is strikingly different from what many family and community members experienced as students themselves. Although many people have achieved personal and professional success and thus consider their own education as adequate preparation for life, we can do better. Too many children are left behind. Teaching practices that focus on helping every student make sense of mathematics have the potential to help achieve the goal of a high-quality mathematics education for every child, not just for those who can learn solely by watching mathematical procedures be demonstrated and explained by an expert. This emphasis does not imply that students should never be expected to listen to or watch others; rather, it underscores the necessity for all students to take an active role in solving, reasoning, and explaining so that they can wrestle with concepts and make sense of procedures under the guidance of an expert who listens to, questions, and challenges their developing ideas. Families and community members who learn about the issues related to mathematics teaching and learning and who support schools in their efforts to achieve a comprehensive mathematics education for every child are the best advocates for the community's children.

To support mathematics teachers and promote mathematics teaching as a profession, family and community members should show their respect and support for teachers and teaching as a profession. If the profession is valued within the community, talented young people are more likely to consider teaching as a career choice. Given the challenging nature of the profession, its crucial importance in helping students develop the mathematical reasoning skills required of adults in an increasingly analytically based world, and the historically high attrition rates associated with the profession, sustained efforts to attract the best and brightest candidates to the field of mathematics teaching are of vital importance. By encouraging schools and school-board members to improve the conditions of teaching, such as competitive salaries and professional development opportunities, family and community members can also have a role in helping keep talented teachers in the profession.

Implementing Change

Teaching is a complex interaction among the teacher, the content being taught, and the students. To make change in the teaching and learning of mathematics, each of these components—teacher, curriculum, and student—must be considered. The total environment in which teaching and learning take place must be reformed.

The vision described in *Principles and Standards* and supported by other NCTM *Standards* documents is still quite different from what is commonly found is most United States and Canadian classrooms. The challenge of changing a familiar educational setting is daunting, but it is necessary in order to improve the learning and performance of students.

The Role of Standards

The recommendations set forth in *Principles and Standards* and its predecessor, *Curriculum and Evaluation Standards for School Mathematics*, were not meant to be interpreted as prescriptions for what must be done at each grade level. They presented a vision of what a high-quality mathematics education for students should entail. This volume provides similar guidance about the kinds of teaching environments, actions, and activities needed to realize the goals for students that were envisioned in previous *Standards* documents. Teachers, administrators, families, other educators, and policymakers in government, business, and industry are expected to work collaboratively to reach consensus on *how* their school mathematics programs need to be changed and *what steps* are needed to make those changes.

Changing School Mathematics

A dialogue on school improvement is ongoing. In local areas, individual schools, districts, and universities are approaching change in different ways and taking steps in different sequences. States have implemented standards for students and teachers, most of which are well aligned with NCTM recommendations. Important to the success of all such efforts is systematic long-term commitment to change that heads in an appropriate direction. One of the strengths of the educational systems in the United States and Canada is their diversity. Different combinations of ideas and strategies will yield many ways to achieve the goal of reforming mathematics teaching and learning. In addition to the possible paths discussed in the NCTM *Standards* documents, the following suggestions focus specifically on teaching and the profession of teaching.

Professionalism

At the present time, teaching as a profession does not receive the public support and esteem that it deserves. Teachers often find themselves in positions in which decisions that have great impact on their ability to teach are being made by persons who do not have the expertise that teachers have gained through their education and experience. Yet the teachers, not the decision makers, are held accountable to the public for the mathematics proficiency of their students. A number of efforts are already under way to rethink the roles of teachers as professionals. The movement to raise teaching to a profession with all the rights and responsibilities entailed is consistent with the vision conceived by the Standards set forth here. For example, the National Board for Professional Teaching Standards' requirements for the certification of teachers of mathematics reflect many aspects of the teaching Standards in this document.

Mathematics teachers entering teaching should have the support of mentors who are experienced teachers of mathematics.

Professional mathematics teachers are accountable for teaching mathematics in an intellectually honest and effective way and for making appropriate instructional decisions. Further, they must be an integral part of the ongoing development and regulation of the profession. Mathematics teachers entering teaching should have the support of mentors who are experienced teachers of mathematics. Not only should programs for the professional development of teachers become an established part of school life, but they should be responsive to teachers' needs at all stages of development. As teachers become more experienced and effective, they should be promoted and accepted as leaders in their schools and in the profession as a whole.

Schools have much to gain by supporting teachers' professional development. The teacher is the key to learning in the classroom. Through the individual acts of teachers, the reform of school mathematics will become a reality. Teachers who have the self-esteem and the intrinsic reward that comes from being encouraged to grow in their profession will provide an environment for students in which students see the excitement and usefulness of mathematics. Further, these teachers will serve as role models to attract talented students to enter the teaching profession.

Structure of schools

With the growth of professionalism comes the need for a fundamental rethinking of the structure of schools. To teach mathematical inquiry, teachers must have an appropriate amount of uninterrupted class time to allow students to explore and discuss ideas. However, implementing extended-length class periods may also raise issues for teachers and administrators to discuss and address. For example, longer class periods should not simply be used as opportunities for students to do their homework in class. The structure of the class period and the time allotted should be matched appropriately. Similarly, allowing students to take a year-long course in the fall of one academic year, then wait until the spring of the following year to take the next mathematics course may lead to year-long absences from mathematics, with concomitant loss of mathematical skills, between consecutive courses.

Mathematics teachers often do not have the necessary resources to do their jobs well.

Additionally, teachers need time to plan, discuss challenges and opportunities with their colleagues, visit one another's classrooms, and engage in curriculum-improvement projects. Such time should be built into the regular school schedule, with additional time set aside in the annual school calendar for in-depth discussions and professional improvement efforts. Some teachers have difficulty obtaining permission and support to attend professional mathematics meetings. Yet through such stimulation, teachers grow and maintain their knowledge about, and enthusiasm for, teaching mathematics and making improvements in classroom instruction. Providing teachers with the support they need to make sound instructional decisions is essential. Mathematics teachers often do not have the necessary resources to do their jobs well. They need calculators, computers, software, manipulatives, and other resources to create the kinds of environments for learning that students need and deserve. Giving teachers more responsibility in budgeting school resources is one effective way of improving mathematics instruction. These kinds of changes, and others perhaps not yet conceived, may provide the stimulus needed to effect real change.

We must think creatively and courageously consider changes in the basic structure of schools; try alternatives and carefully study their effects; and create different working models of school structures in which students' mathematical proficiency and teachers' professional growth far exceed those in today's models.

Entry into the profession

The current shortage of qualified mathematics teachers and the changing nature of the pool from which teachers come have spurred experimentation with different forms of initial licensure. Alternative certification programs that offer a form of in-school apprenticeship to persons holding undergraduate subject-matter degrees are being tried in various places. Other forms of initial licensure for teachers are also being used in some states. *Mathematics Teaching Today* offers guidance for developers of induction and licensure programs. Whether teachers enter teaching through four-year, five-year, or alternative certification programs, it is essential that they know the subject matter of mathematics, how students think about mathematics, strategies for teaching mathematics, how to select or create mathematical tasks, and how to create an environment for learning mathematics in which *all* students have an opportunity to experience a comprehensive mathematics education.

Teachers of early childhood, elementary, middle, and high school mathematics need broad and deep knowledge in three fundamental domains: mathematics, mathematics teaching, and students. This knowledge should be appropriate to the level that they will teach or are teaching. However, teachers need to know both where their students have been and where their students are going mathematically. This recommendation means that an early childhood or elementary school teacher needs to have experience with the big ideas of school mathematics at all levels. In addition, the high school teacher needs to understand what significant mathematical ideas are taught prior to high school and how those ideas are represented. Middle school teachers need to be able to connect what they are teaching with the elementary school experiences of students and also to anticipate the growth of mathematical ideas as the students proceed into high school. Therefore, all programs need to be examined carefully to determine whether they help teachers in the program develop a broad view of the mathematics curriculum, student learning, and teaching beyond that which is required for their own grade level.

In addition to strong preparation programs, all teachers should know that the profession of teaching involves lifelong learning. Regardless of how state certification programs are structured, the profession demands that teachers continue to learn about mathematics and about teaching mathematics throughout their careers.

School mathematics as a part of mathematics

Mathematics has changed dynamically since the 1960s, but school mathematics, for the most part, has not reflected those changes. It is not perceived as having much connection with "real" mathematics, even by teacher education students at colleges and universities. Part of the responsibility for this schism between school mathematics and the mathematics studied at the university level rests on the shoulders of university and college mathematicians. Preservice teachers seldom have opportunities to see how the mathematics they are studying relates to the mathematics of schools. In addition, students at both levels, school and collegiate, are often taught an outmoded curriculum that has very little to do with what is important in mathematics today. A reexamination of the relationship of school mathematics to the discipline of mathematics is a necessary part of the needed reform.

Collegiate curriculum

Teachers need to learn in technology-rich environments if they are to teach using technology.

Principles and Standards and the Standards in this volume articulate a vision to guide the reform of school mathematics curriculum, teaching, and evaluation. However, if teachers are to change the way they teach, they need to learn significant mathematics in situations in which good teaching is modeled. Some institutions involved in teacher preparation have explored and modified aspects of the undergraduate curriculum. We believe that this effort should continue and be expanded to include consideration of the entire undergraduate curriculum and, perhaps even more important, the models of instruction used in collegiate classrooms. For example, technology and its use in doing, teaching, and learning mathematics is a responsibility of the mathematics community as well as the mathematics education community. Teachers need to learn in technology-rich environments if they are to teach using technology.

Collaboration between schools and universities

The creation of new structures for the professional development of teachers should help blur the lines between universities and schools. The interaction of university faculty and school faculty as colleagues with different areas of expertise is likely to improve the teaching and learning of mathematics at all levels. Mathematicians have a responsibility to find creative ways to share the excitement of new advances in mathematics both with schoolteachers and with their students. *The pool of young people who are interested in pursuing professions in mathematics or the sciences is far too small to meet the needs of society.* From this pool of young people who have an interest in mathematics will come the teachers of tomorrow. Raising the prestige and rewards of teaching is crucial to attracting talented and caring young people into teaching.

Mathematics educators at all levels need to take responsibility for working together to promote all students' interest in mathematics

Of particular concern is the small representation in the scientific pool—and hence in the pool of teachers—of women, ethnic minorities, physically challenged, and other underrepresented groups. Mathematicians and mathematics educators at all levels have a responsibility to invest time, energy, and their creative talents in finding ways to communicate the excitement and usefulness of mathematics to young people, and to devise programs that help underrepresented students succeed in their study of mathematics. Beginning these efforts at the university level is too late. To have the desired impact, we must begin at the elementary school level. Mathematics educators at all levels need to take responsibility for working together to promote all students' interest in mathematics in elementary school and to help maintain that interest through middle school, high school, and beyond.

As schools and universities strengthen their ongoing communication, educators can articulate mathematics programs across elementary school, middle school, high school, and college levels. Administrators and others in leadership positions can establish support groups for teachers at all levels who are attempting to implement the Standards.

Accrediting and certificating agencies

Just as tests are influential in determining the mathematics curriculum, accrediting agencies such as the National Council for the Accreditation of Teacher Education, The Teacher Education Accreditation Council, and state or provincial certificating agencies influence collegiate teacher preparation programs. These agencies can be a constraint on needed change or a force for the improvement of teaching. Such agencies can play a positive role in reform as they work with professional organizations and schools to be sure that their guidelines are in tune with the goals and vision of teaching that the profession espouses.

For example, accrediting agencies can set guidelines that expect schools of education and mathematics departments to be technology-rich environments in which to learn. The NCTM Standards documents represent the consensus of the mathematics and mathematics education communities and can guide accrediting agencies in determining their criteria for judging mathematics teacher preparation programs.

States and provinces have departments of education that are responsible for monitoring school programs, requirements, and offerings. They also are the principal agents for the initial certification of teachers. In both of these roles, state departments of education should review their practices in light of the recommendations in the NCTM *Standards* documents. An additional role for departments of education is to ensure that professional development opportunities are organized and available to help teachers and schools implement these Standards for school mathematics curricula and teaching.

Networking with other disciplines

An interesting aspect of recent efforts toward school improvement is that different disciplines are discovering common goals and common strategies for achieving them. For instance, language arts teachers are working on communication, which is a major goal for the mathematics curriculum at all levels. Networking, sharing ideas, learning from one another, and helping support another discipline in our mathematics classrooms contribute to everyone's success. Students can see the commonality of strategies for attacking problems and the benefit of discussion and argumentation in refining ideas. They should also see these commonalities in social studies, science, language arts, and all school subjects. Cross-discipline studies give additional meaning to strategies and concepts and add interest and variety to the learning process. Given the few hours in a school day and the amount of material to be covered, collaborative efforts to seek common ground and offer mutual support serve the best interest of teachers in all disciplines.

Research

The vision set forth in this volume is based on a set of values and beliefs about mathematics teaching and learning that are consistent with current research. However, these Standards suggest a research agenda with respect to teacher education and learning to teach. Much that we need to know cannot be determined from current practice. We need experimentation and careful research, new structures of schools, new interactions between universities and schools, new teacher education programs, school and university professional development programs, teaching and learning with computing technology and other forms of technology and tools, new forms of instruction in university and school mathematics classes, and other aspects of reform. Researchers and some teachers are already engaged in accumulating evidence in many of these areas (see, e.g., Mewborn 2006). A continuation of these efforts is vital, and universities, schools, and the mathematics education community must value such research. Results of such studies are needed to guide us on the many possible paths to reform in mathematics teaching and learning.

The kind of teaching envisioned in these Standards will take time to become reality. We need to understand the trade-offs of changing from "covering" a broad set of mathematical topics to conducting more focused, in-depth investigations of fewer mathematical ideas (NCTM 2006). We need to understand better how to meet the mathematical needs of a diverse student population. We need to understand better how participation in small groups and classroom discourse can be used to help students learn to make mathematical judgments and enhance their conceptual understanding of mathematics. The instructional use of manipulatives, calculators, computers, and other tools and technologies for teaching mathematics needs to be continually studied, with a particular concern for the needs of diverse students. As reform proceeds, many other issues will arise that will need careful study. Research is yet another arena in which schools and universities have much to gain by collaborating with each other.

Summary

Many possible next steps can be taken to improve mathematics teaching and learning. If we make a long-term commitment to the Standards set forth in this document, in *Principles and Standards*, and in other NCTM *Standards* documents; if we approach the task with the will to persevere; if we are critical of the steps we take; and if we make appropriate midcourse corrections, we will make progress toward the goal of more and better mathematics for all students. The picture of mathematics teaching and learning envisioned in these Standards is ambitious. We will not reach this goal overnight. Such change will require much work and dedication from teachers and many others. However, this effort is essential if we are to improve mathematics learning for our students. We must be impatient enough to take action and patient enough to sustain our efforts until we see results.

We must be impatient enough to take action and patient enough to sustain our efforts until we see results.

We urge you to start by reading *Principles and Standards for School Mathematics* as well as this document. Talk to your colleagues. Discuss these ideas with families; school and university administrators; and others in government, business, and industry. Collectively and individually set goals for change, then establish a plan that will guide change over the next several years. Seek resources to support that plan. Be a part of working to make a comprehensive mathematics education a reality for every student.

Chapter 5

Questions for the Reflective Practitioner

In this chapter, we present questions for reflection that are based on recommendations made in the preceding chapters of *Mathematics Teaching Today: Improving Practice, Improving Student Learning.* These questions were designed to inspire thoughtful reflection and discussion by individuals or within groups of preservice teachers, practicing teachers, teacher leaders, department chairs, principals, district-level administrators, teacher educators, professional developers, parents, community members, and others with an interest in, and commitment to, improving the teaching and learning of school mathematics.

The questions are presented along with the Standards to which they are intended to relate. The questions have no right or wrong answers; rather, they are designed to prompt readers to consider the issues from their personal and professional perspectives. Everyone contributes to the teaching and learning of mathematics, either directly or indirectly through involvement with students, teachers, or schools. The challenge for each individual is to figure out how he or she can contribute to the collective effort to realize the vision of more and better mathematics for all students.

Discussion Questions on Standards for the Teaching and Learning of Mathematics

The questions in this section are written as if they are being posed to classroom teachers. Others also are encouraged to consider these questions from their respective professional perspectives. Readers also may ask themselves, "How can I support a teacher who is being asked this question?"

Questions on Standard 1—Knowledge of Mathematics and General Pedagogy

1. If you could enhance your content knowledge in one area of mathematics, what would it be? What resources would you use to learn more about this content area? How would you assess the growth of your own understanding?
2. If you could enhance your knowledge in one aspect of pedagogy, what would it be? What resources would you use to learn more about that aspect of pedagogy? How would you assess the growth of your own understanding?
3. If you could enhance your knowledge of one method of assessment, what method would it be? What resources would you use to learn more about that method of assessment? How would you assess the growth of your own understanding?

Questions on Standard 2—Knowledge of Student Mathematical Learning

1. What specific practical steps you can take to increase your students' active involvement during mathematics class?
2. In what ways would you like to see your students' mathematical communication skills enhanced? What instructional method or methods can you use that could help your students see the need to improve their mathematical communication?
3. Consider a topic that you plan to teach in the next few weeks. Identify the main concepts and procedures associated with that topic. How will your instruction and assessment vary when your instructional focus is conceptual understanding compared with procedural understanding?

Questions on Standard 3—Worthwhile Mathematical Tasks

1. Name one connection that you want your students to make between two mathematical content areas. How do your current teaching materials lend themselves to helping students make that connection? What resources can you tap for supplementary materials that may help you better achieve your goal?
2. What set of tools and heuristics (strategies) have your students developed to help them solve unfamiliar problems? Find a rich, open-ended problem that can be approached using various strategies. Challenge your students to solve the problem in as many ways as possible. Analyze the results of this activity. Did your students use as many methods as you would like? How can you enhance their problem-solving repertoire?
3. Analyze the teaching materials you use. In what ways do those materials draw on students' diverse backgrounds and experiences? Can you identify any bias in the teaching materials you use? What types of tasks are the most inclusive in terms of culture, gender, or other demographic features?

Questions on Standard 4—Learning Environment

1. Reflect on your daily classroom routine. To what activities do you devote the most time? To what activities do you devote the least time? What messages might your classroom routines send to students about what is valued in your classroom? What messages might those routines send about the nature of mathematics?
2. What role does technology play in your classroom? How do students decide when technology would be an appropriate tool to use? What technological tools would you like your students to use more regularly? What technological tools would you like to learn more about yourself?

3. Consider your students' motivations for completing mathematical tasks. Are they focused on finding answers or on making sense of mathematical ideas? What can you do to encourage students to reflect on their own thinking as well as on mathematical structures?
4. How do you encourage your students to take intellectual risks? In what ways do you let students know that although their classroom contributions will be critiqued by the classroom community of learners, the focus of those discussions will be on the value of the reasoning rather than on the value of the person doing the reasoning?
5. In what ways can you help students adapt to an environment that is predominantly student centered rather than teacher centered?

Questions on Standard 5—Discourse

1. What factors do you consider when assessing the quality of classroom discourse? How do you jump-start a stalled or dead-end discussion or presentation?
2. In what ways do you encourage students to carefully listen to one another's ideas?
3. What types of questions or prompts have you found to be most effective for getting students to explain their thinking and their solutions?

Questions on Standard 6—Reflection on Student Learning

1. What level of responsibility can or should your students take for their own learning?
2. What means do you use to assess students' understanding on a daily basis?
3. How do you ensure consistency, in form and expectations, between your instruction and your assessment of students?
4. What different types of assessments do you use? What is the purpose of each type?
5. What role can group assessment or self-assessment play in your classes?

Questions on Standard 7—Reflection on Teaching Practice

1. Have you or your students been surprised by the results of a recent quiz or test? What can you do to ensure that no one in your class is surprised by assessment results?
2. What instructional challenges do you face that you would like to find ways to address?

3. What means can you use to determine whether instructional modifications are successful?
4. Given time and resources, what would you most like to have a colleague do to help you reflect on and improve your teaching?

Discussion Questions on Standards for the Observation, Supervision, and Improvement of Mathematics Teaching

The questions in this section are written as if they are being posed to teachers and supervisors of classroom teachers. Others also are encouraged to consider these questions from their respective professional perspectives. Readers may also ask themselves, "How can I support teachers or supervisors who are being asked to consider these issues?"

Questions on Standard 1—The Continuous Improvement Cycle

1. What kinds of reflection and analysis do you engage in before the school year begins for the purpose of setting goals for improvement in the coming year?
2. In what ways do you work with your supervisors, teachers, or colleagues to analyze teaching and learning?
3. How do you identify areas for improvement, and how do you measure success in addressing those areas?

Questions on Standard 2—Teachers as Participants in the Observation, Supervision, and Improvement Process

1. In what ways do you or the teachers you mentor actively contribute to the assessment and improvement of your respective professional practice?
2. Who sets goals for improvement for each school year? Who chooses the focus of, or activities for, professional days?
3. If you had no time or monetary constraints, what would be your ideal professional growth experience? Why? What could you do to help make that ideal a reality?
4. How can you encourage and support collaborative planning and decision making to enhance teachers' professional growth and, ultimately, children's learning?

Questions on Standard 3—Data Sources for the Observation, Supervision, and Improvement of Mathematics Teaching

1. What are the benefits and limitations of using lesson plans to monitor your performance as a teacher or the performance of the teachers you supervise? What other information would be useful for documenting the careful planning and thoughtful implementation of each lesson?
2. What are the benefits and limitations of using student work samples to assess the effectiveness of specific lessons or units? What other information would be useful for documenting the quality of student learning?
3. What other sources of data would you want to include in the periodic assessment of teaching and learning? Why?
4. In what ways can you facilitate and support collaborative efforts of teachers to enhance teaching and learning?

Questions on Standard 4—Teacher Knowledge and Implementation of Worthwhile Mathematical Tasks

1. What types of tasks would indicate to a supervisor and to students that a teacher values proof and reasoning as fundamental aspects of mathematics?
2. In what ways can you tell whether students are engaged in genuine problem solving rather than simply completing routine exercises?
3. In what ways can you encourage teachers or students to incorporate a variety of mathematical representations into tasks and their solutions?
4. What are some of the richest tasks you have used or seen used in a mathematics classroom? What characteristics made them so mathematically worthwhile?

Questions on Standard 5—Teacher Knowledge and Implementation of Effective Learning Environments and Mathematical Discourse

1. If students are to be encouraged to struggle with, and make sense of, mathematical ideas, how should that focus be reflected in the classroom discourse? What portion of the class time should include students' expressing their mathematical ideas in words or in writing?
2. What type of questions do you use or do you look for to ensure that students are actively engaged in making connections within mathematics or between mathematics and other disciplines?
3. What types of conjectures do your students make? What do such conjectures indicate about students' understanding of mathematical concepts?

4. In what settings and activities do your students engage in inductive reasoning or deductive reasoning? What types of situations call for each type of reasoning?
5. In what ways are all students in the classroom being challenged with mathematical tasks that maintain their interest and help them see the world from a mathematical perspective? What obstacles might prevent you from holding high expectations for all students? How can teachers and supervisors work together to overcome those obstacles?

Questions on Standard 6—Teacher Knowledge and Implementation of Assessment for Students' Mathematical Understanding

1. Examine some of the assessments that you and your colleagues use on a regular basis. In what ways might inherent biases be reflected in those assessments? How can you or your colleagues ensure greater equity in student assessment practices?
2. What mathematical concepts and processes are the most difficult to assess? What makes assessing them so difficult? What types of evidence would convince you that a student has a good grasp of one of those concepts or processes?
3. What nontraditional assessment method would you like to learn more about or implement more often? What types of information about student understanding might this method generate that would not be as easy to uncover using more traditional timed quizzes or tests?
4. What portion of your classroom assessments are formative, leaving room for student or teacher improvement on the basis of assessment results? What actions are students expected to take in response to assessment results? What actions are teachers expected to take in response to assessment results?

Discussion Questions on Standards for the Education and Continued Professional Growth of Teachers of Mathematics

The questions in this section are written as if they are being posed to university teacher educators, professional developers, mathematicians, and others involved in the education of teachers. Others also are encouraged to consider these questions from their respective professional perspectives.

Questions on Standard 1—Teachers' Mathematical Learning Experiences

1. What role can you play in fostering productive collegial interactions among mathematicians, university teacher educators, and professional developers in the interest of providing the best possible mathematical learning experiences for teacher candidates and teachers?
2. In your role as an educator of preservice teachers or practicing teachers, what can you do to model such practices as those described in this and other NCTM *Standards* documents? In what ways can you help the preservice teachers or practicing teachers you work with understand why you are using those instructional methods?

Questions on Standard 2—Knowledge of Mathematical Content

1. In what ways are preservice teachers' or practicing teachers' mathematical learning experiences aligned with the mathematical learning experiences they are expected to provide for their school students?
2. What role does problem solving play in basic content courses for teacher candidates or teachers? How are teacher candidates or teachers prepared to teach heuristics and explore multiple solution methods with their own students?
3. How are conjecturing, problem posing, reasoning, and proof incorporated into your content courses for teacher candidates or teachers?
4. In what ways can connections among content courses, between content courses and the prekindergarten–grade 12 curriculum, or between mathematics and other disciplines be enhanced in the mathematical learning experiences of teacher candidates or teachers?
5. In what ways do the content-learning experiences in your teacher-preparation or teacher-enhancement program help develop the robust and connected mathematical understandings needed for teaching?

Questions on Standard 3—Knowledge of Students as Learners of Mathematics

1. Given the great diversity in society, what kinds of experiences can help teachers understand, celebrate, and draw on the gender, cultural, ethnic, socioeconomic, and other differences in their classrooms?
2. What do teachers need to know to be able to adequately assess students' mathematical thinking, encourage students' active participation, and promote opportunities to learn mathematics for all students?
3. In what ways can your teacher-development or teacher-enhancement program help teachers see connections between research and practice? What sorts of

action research opportunities would you like to make available to teacher candidates or teachers with whom you work?

4. What do teachers need to know about learners with special needs to be well prepared to make appropriate accommodations for those students? In what ways should this expertise be developed?

Questions on Standard 4—Knowledge of Mathematical Pedagogy

1. What role should technology have in the mathematical preparation or enhancement of teachers? In what ways are teacher candidates or teachers with whom you work being prepared to use current technology in their classrooms and to continue to learn about technology in the future?
2. What kinds of experiences can help teachers learn to effectively facilitate classroom discussions? What do teachers need to know to learn to be a guide on the side rather than a sage on the stage?
3. Why is building learning communities within their classrooms an important goal for teachers? What is the nature of mathematical authority in a learning community? What methods can teachers use to determine whether the norms and practices in their classroom are conducive to reasoning, argumentation, and justification?
4. How can teachers learn how various instructional strategies influence student understanding? How can they develop and improve their repertoire of strategies?

Questions on Standard 5—Participation in Career-long Professional Growth

1. What are some of the essential components of effective professional growth opportunities for teachers of mathematics?
2. How often should teachers engage in professional growth experiences?
3. Who should decide what kind of experiences are appropriate for promoting teachers' professional growth?
4. What would you like to know more about in terms of teaching mathematics or preparing teachers? What types of professional growth opportunities would be the most beneficial for mathematics teacher educators, mathematicians, or others involved in teacher development?
5. In what settings are you asked to share your perspectives about mathematics teaching and learning? What information can you share in those settings that might help others appreciate the complexity of the job of teaching mathematics and the expertise that teachers must bring to that role?

Appendix

Vignettes—Alignment with *Principles and Standards for School Mathematics*

Vignettes, Chapter 1—"Standards for Teaching and Learning Mathematics": Alignment with *Principles and Standards for School Mathematics*

Vign. No.	Vignette Title	Principle	Content Standard	Process Standard	Grade Band
1.1	Drawing on Mathematical Knowledge during Exploration Activities	Teaching, Learning, Technology	Number and Operations	Reasoning and Proof	Grades 3–5
1.2	Using Student Interviews to Identify Misconceptions	Teaching, Learning, Assessment	Number and Operations	Communication, Representation	Pre-K–2
2.1	Modifying Resources to Meet Students' Needs	Curriculum, Learning	Number and Operations	Representation, Connections	Grades 6–8
2.2	Helping Students Build on Informal Understandings	Curriculum, Teaching, Learning	Geometry	Communication	Grades 6–8
3.1	Selecting Intellectually Stimulating Tasks	Equity, Curriculum, Teaching	Measurement	Problem Solving, Reasoning and Proof, Connections	Grades 3–5
3.2	Making Sense of Algebraic Expressions	Teaching, Assessment	Algebra	Communication, Representation	Grades 9–12
4.1	The Role of the Environment in Supporting Student Development	Equity, Assessment, Technology	Algebra	Communication, Connections, Representation	Grades 9–12
4.2	Encouraging Sense-Making by Expecting Students to Reason	Equity, Learning	Number and Operations	Problem Solving, Reasoning and Proof, Communication, Representation	Pre-K–2
5.1	Only the Nose Knows, but the Children Can Reason!	Equity, Teaching	Number and Operations	Reasoning and Proof, Communication	Pre-K–2
5.2	Making the Transition from Student-Invented to Standard Representations	Curriculum	Number and Operations	Reasoning and Proof, Representation	Pre-K–2
5.3	Letting the Discourse Happen: Monitoring Collaborative Groups	Equity, Teaching, Learning, Technology	Data Analysis and Probability	Communication, Reasoning and Proof	Grades 9–12
6.1	Using a Challenging Problem to Assess Student Understanding	Assessment, Technology	Algebra	Problem Solving, Communication, Connections	Grades 9–12
6.2	Documenting Student Work to Monitor and Report Progress	Assessment	Number and Operations	Reasoning and Proof, Communication, Connections	Grades 3–5
7.1	Examining Interaction Patterns in the Classroom	Equity	All	Communication	All

Vignettes, Chapter 2—"Standards for the Observation, Supervision and Improvement of Mathematics Teaching": Alignment with *Principles and Standards for School Mathematics*

Vign. No.	Vignette Title	Principle	Content Standard	Process Standard	Grade Band
1.1	The Homework Dilemma	Equity, Assessment	Algebra	Problem Solving	Grades 9–12
2.1	Using Technology to Enhance Learning	Curriculum, Learning, Technology	Data Analysis and Probability	Reasoning and Proof, Communication, Connections	Grades 6–8
3.1	Multiple Episodes of Classroom Teaching	Equity, Learning, Assessment	Number and Operations, Algebra	Representation, Connections	Grades 6–8
3.2	The Power of Lesson Study	Equity, Curriculum, Teaching, Learning, Assessment	Algebra	Communication	Grades 9–12
4.1	Multiple Representations for Number and Operations	Curriculum, Teaching, Learning, Assessment	Number and Operations, Measurement	Communication, Connections, Representation	Grades 3–5
4.2	Real-Life Connections	Teaching, Assessment, Technology	Algebra, Measurement, Data Analysis and Probability	Connections	Grades 9–12
5.1	Student Involvement in Classroom Discourse	Teaching, Assessment	Algebra	Reasoning and Proof, Communication	PreK–2
5.2	Making Mathematical Conjectures	Teaching, Assessment, Technology	Algebra	Reasoning and Proof, Communication	Grades 9–12
6.1	Assessing Student Understanding	Assessment	Number and Operations	Connections, Representation	PreK–2
6.2	Using a Variety of Assessment Methods	Equity, Teaching, Assessment	Algebra, Geometry	Problem Solving, Reasoning and Proof, Communication	Grades 9–12

Vignettes, Chapter 3—"Standards for the Education and Continued Professional Growth of Teachers of Mathematics": Alignment with *Principles and Standards for School Mathematics*

Vign. No.	Vignette Title	Principle	Content Standard	Process Standard	Grade Band
1.1	Changing Learning Expectations: An Instructor's Role	Learning and Technology	Data Analysis and Probability	Problem Solving, Reasoning and Proof, Connections	Grades 6–8, Grades 9–12
1.2	Modeling Instructional Strategies: Using Technology in Calculus	Teaching, Technology	Algebra	Communication Representation	Grades 6–8, Grades 9–12
2.1	Unlocking the Locker Problem	Learning	Number and Operations	Problem Solving, Connections	All
2.2	Connecting Content to the Secondary Curriculum: Behavior of Functions	Curriculum, Teaching, Learning	Algebra	Connections	Grades 9–12
3.1	Assessing and Building on Students' Fraction Concepts	Equity, Teaching, Learning, Assessment	Number and Operations	Problem Solving, Reasoning and Proof, Communication	Grades 3–5
3.2	Exploring Statistical Thinking: An Interview with Sara	Assessment, Teaching	Data Analysis and Probability	Problem Solving, Communication, Representation	Grades 6–8
3.3	Facing a Teaching Dilemma: Gender Interaction	Equity, Teaching, Assessment	All	Reasoning and Proof, Communication	Grades 3–5
4.1	Considering Assessment Strategies: The Interview	Assessment	Number and Operations	Communication, Representation	All
4.2	Learning to Recognize and Exploit Multiple Solutions Methods	Teaching, Learning, Assessment	Number and Operations, Algebra	Problem Solving, Communications	Grades 9–12
5.1	Rethinking Teaching Strategies: The Condominium Problem	Curriculum, Teaching	Number and Operations	Problem Solving, Reasoning and Proof, Representation	All
5.2	Redesigning Instruction: Collegial Support	Equity, Teaching, Learning	Algebra	Reasoning and Proof, Representation	Grades 3–5, Grades 6–8
5.3	Expanding Perspectives on Technology	Technology, Assessment	Algebra, Geometry	Connections, Representation	Grades 9–12

Bibliography

American Mathematical Society NCTM2000 Association Resource Group. "Reports of the AMS Association Resource Group." *Notices of the American Mathematical Society* 45 (February 1998): 270–76. Also available online at http://www.ams.org/notices/199802/comm-amsarg.pdf (accessed June 19, 2006).

Artzt, Alice F., and Eleanor Armour-Thomas. *Becoming a Reflective Practitioner: A Guide for Observations and Self-Assessment.* Mahwah, N.J.: Lawrence Erlbaum Associates, 2002.

Ball, Deborah L., and Hyman Bass. "Interweaving Content and Pedagogy in Teaching and Learning to Teach: Knowing and Using Mathematics." In *Multiple Perspectives on the Teaching and Learning of Mathematics,* edited by Jo Boaler, pp. 83–104. Westport, Conn.: Ablex Publishing Corporation, 2000a.

———."Making Believe: The Collective Construction of Public Mathematical Knowledge in the Elementary Classroom." In *Yearbook of the National Society for the Study of Education, Constructivism in Education,* edited by Denis C. Phillips, pp. 193–224. Chicago: University of Chicago Press, 2000b.

Ball, Deborah L., and David K. Cohen. "Developing Practice, Developing Practitioners: Toward a Practice-Based Theory of Professional Education." In *Teaching as the Learning Profession,* edited by Linda Darling-Hammond and Gary Sykes, pp. 3–32. San Francisco: Jossey-Bass, 1999.

Ball, Deborah L., Heather C. Hill, and Hyman Bass. "Knowing Mathematics for Teaching: Who Knows Mathematics Well Enough to Teach Third Grade, and How Can We Decide?" *American Educator* (Fall 2005): 14–46. Also available online at http://www.aft.org/pubs-reports/american_educator/issues/fall2005/index.htm (accessed June 19, 2006).

Ball, Deborah L., Sarah T. Lubienski, and Denise S. Mewborn. "Research on Teaching Mathematics: The Unsolved Problem of Teachers' Mathematical Knowledge." In *Handbook of Research on Teaching,* 4th ed., edited by Virginia Richardson, pp. 433–56. New York: Macmillan, 2001.

Berliner, David C. "Our Impoverished View of Educational Reform." *Teachers College Record* 108, no. 6 (2006): 949–95. http://www.tcrecord.org ID Number: 12106 (accessed: June 21, 2006).

Borko, Hilda, and Carol Livingston, "Cognition and Improvisation: Differences in Mathematics Instruction by Expert and Novice Teachers." *American Educational Research Journal* 26 (Winter 1989): 473–98.

Cobb, Paul, and Erna Yackel. "Constructivist, Emergent, and Sociocultural Perspectives in the Context of Developmental Research." *Educational Psychologist* 31 (Summer/Fall 1996): 175–90.

Conference Board of the Mathematical Sciences. *The Mathematical Education of Teachers*. Providence, R.I., and Washington, D.C.: American Mathematical Society and Mathematical Association of America, 2001. Also available online at http://www.cbmsweb.org/MET_Document/index.htm (accessed June 19, 2006).

Confrey, Jere. "What Constructivism Implies for Teaching." In *Constructivist Views on the Learning and Teaching of Mathematics,* Monograph No. 4 of the *Journal for Research in Mathematics Education,* edited by Robert B. Davis, Carolyn A. Maher, and Nel Noddings, pp. 107–22. Reston, Va.: National Council of Teachers of Mathematics, 1990.

Cuoco, Al, E. Paul Goldenberg, and June Mark. "Habits of Mind: An Organizing Principle for Mathematics Curriculum." *Journal of Mathematical Behavior* 15 (December 1996): 375–402.

DuFour, Richard. "What Is a Professional Learning Community?" *Educational Leadership* 61 (May 2004): 6–11.

DuFour, Richard, and Robert Eaker. *Professional Learning Communities at Work: Best Practices for Enhancing Student Achievement.* Bloomington, Ind.: National Educational Service, 1998.

Farberman, Rhea K. "Immigration's Impact: Experts at an APA Summit Explored How Immigration Is Driving America's Changing Demographics—and Will Change the Nation." *Monitor on Psychology* 37 (March 2006): 42.

Fernandez, Clea, and Makoto Yoshida. *Lesson Study: A Japanese Approach to Improving Mathematics Teaching and Learning.* Mahwah, N.J.: Lawrence Erlbaum Associates, 2004.

Ferrini-Mundy, Joan, and W. Gary Martin. "Using Research in Policy Development: The Case of the National Council of Teachers of Mathematics' *Principles and Standards for School Mathematics*." In *A Research Companion to* Principles and Standards for School Mathematics, edited by Jeremy Kilpatrick, W. Gary Martin, and Deborah Schifter, pp. 395–413. Reston, Va.: National Council of Teachers of Mathematics, 2003.

Fox, Lynn, and Janet Soller. "Psychosocial Dimensions of Gender Differences in Mathematics." In *Perspectives on Gender,* edited by Judith E. Jacobs, Joanne Rossi Becker, and Gloria F. Gilmer, pp. 9–24. Reston, Va.: National Council of Teachers of Mathematics, 2001.

Fullan, Michael. "Professional Communities Writ Large." In *On Common Ground,* edited by Richard DuFour, Robert Eaker, and Rebecca DuFour, pp. 209–223. Bloomington, Indiana: National Education Service, 2005.

Fuson, Karen C., William M. Carroll, and Jane V. Drueck. "Achievement Results for Second and Third Graders Using the Standards-Based Curriculum Everyday Mathematics." *Journal for Research in Mathematics Education* 31 (May 2000): 277–95.

Gilligan, Carol. *Making Connections: The Relational Worlds of Adolescent Girls at Emma Willard School.* Cambridge, Mass.: Harvard University Press, 1990.

Goldin, Gerald A. "Epistemology, Constructivism, and Discovery." In *Constructivist Views of the Teaching and Learning of Mathematics,* edited by Robert B. Davis, Carolyn A. Maher, and Nel Noddings. Reston, Va.: National Council of Teachers of Mathematics, 1990.

Henningsen, Marjorie, and Mary K. Stein. "Mathematical Tasks and Student Cognition: Classroom-Based Factors That Support and Inhibit High-Level Mathematical Thinking and Reasoning." *Journal for Research in Mathematics Education* 28 (November 1997): 524–49.

Hiebert, James. "What Research Says about the NCTM Standards." In *A Research Companion to* Principles and Standards for School Mathematics, edited by Jeremy Kilpatrick, W. Gary Martin, and Deborah Schifter, pp. 5–23. Reston, Va.: National Council of Teachers of Mathematics, 2003.

———. "Relationships between Research and the NCTM Standards." *Journal for Research in Mathematics Education* 30 (January 1999): 3–19.

Hiebert, James, Thomas P. Carpenter, Elizabeth Fennema, Karen C. Fuson, Diana Wearne, Hanlie Murray, Alwyn Olivier, and Piet Human. *Making Sense: Teaching and Learning Mathematics with Understanding.* Portsmouth, N.H.: Heinemann, 1997.

Hiebert. James, Ronald Gallimore, Helen Garnier, Karen Bogard Givvin, Hilary Hollingsworth, Jennifer Jacobs, Angel Miu-Ying Chui, Diana Wearne, Margaret Smith, Nicole Kersting, Alfred Manaster, Ellen Tseng, Wallace Etterbeek, Carl Manaster, Patrick Gonzales, and James Stigler. *Highlights from the TIMSS 1999 Video Study of Eighth-Grade Mathematics Teaching.* NCES 2003-011. Washington, D.C.: U.S. Department of Education, National Center for Education Statistics, 2003a. Also available online at http://nces.ed.gov/pubsearch/pubsinfo.asp?pubid=2003011 (accessed June 28, 2006).

———. *Teaching Mathematics in Seven Countries: Results from the TIMSS 1999 Video Study.* Washington, D.C.: National Center for Education Statistics, 2003b. Also available online at http://nces.ed.gov/pubsearch/pubsinfo.asp?pubid=2003013 (accessed June 28, 2006).

Hiebert, James, Ronald Gallimore, and James W. Stigler. "The New Heroes of Teaching." *Education Week* 23, no. 10 (November 5, 2003): 56, 42. Also available online at http://www.edweek.org/ew/ewstory.cfm?slug=10hiebert.h23 and http://www.lessonlab.com/about/index.cfm/c/ab_84 (accessed June 21, 2006).

Hiebert, James, and Diana Wearne. "Developing Understanding through Problem Solving." In *Teaching Mathematics through Problem Solving, Grades 6–12,* edited by Harold L. Schoen, pp. 3–14. Reston, Va.: National Council of Teachers of Mathematics, 2003.

Hill, Heather C., Brian Rowan, and Deborah L. Ball. "Effects of Teachers' Mathematical Knowledge for Teaching on Student Achievement." *American Educational Research Journal* 42 (Summer 2005): 371–406.

Hill, Heather C., Stephen G. Schilling, and Deborah L. Ball. "Developing Measures of Teachers' Mathematics Knowledge for Teaching." *Elementary School Journal* 105, no. 1 (September 2004): 11–30.

Howard, Keith. "Stereotypes as Threats on Tests." *USC Urban Ed: The Magazine of the USC Rossier School of Education* (Winter/Fall 2004–2005): 41–43. Also available online at http://www.usc.edu/dept/education/news/urbaned04_05/40.pdf (accessed June 21, 2006).

Jacobs, Judith, and Joanne Becker. "Introduction." In *Perspectives on Gender,* edited by Judith Jacobs and Joanne Becker, pp. 1–7. Reston, Va.: National Council of Teachers of Mathematics, 2001.

Khisty, Lena L. "Making Inequality: Issues of Teachers' Language Use in the Instruction of Mathematics." In *New Directions for Equity in Mathematics Education,* edited by Walter G. Secada, Elizabeth Fennema, and Lisa B. Adajian. New York: Cambridge University Press, 1995.

Kilpatrick, Jeremy, W. Gary Martin, and Deborah Schifter, eds. *A Research Companion to* Principles and Standards for School Mathematics. Reston, Va.: National Council of Teachers of Mathematics, 2003.

Konold, Clifford, and Traci Higgins. "Reasoning about Data." In *A Research Companion to* Principles and Standards for School Mathematics, edited by Jeremy Kilpatrick, W. Gary Martin, and Deborah Schifter, pp. 5–23. Reston, Va.: National Council of Teachers of Mathematics, 2003.

Kulik, James A. *An Analysis of the Research on Ability Grouping: Historical and Contemporary Perspectives*. Storrs, Conn.: National Research Center on the Gifted and Talented, University of Connecticut, 1992. RBDM 9204. Summary available at http://www.gifted.uconn.edu/nrcgt/kulik.html (accessed June 14, 2005).

Lampert, Magdalene. *Teaching Problems and the Problems of Teaching*. New Haven, Conn.: Yale University Press, 2001.

Lappan, Glenda, and Mary K. Bouck. "Developing Algorithms for Adding and Subtracting Fractions." In *The Teaching and Learning of Algorithms in School Mathematics: 1998 Yearbook,* edited by Lorna J. Morrow, pp. 183–97. Reston, Va.: National Council of Teachers of Mathematics, 1998.

Lavasser, Kenneth, and Al Cuoco. "Mathematical Habits of Mind." In *Teaching Mathematics through Problem Solving: Grades 6-12,* edited by Harold L. Schoen (Volume Editor) and Randall I. Charles (Series Editor), pp. 27–37. Reston, Va.: National Council of Teachers of Mathematics, 2003.

Leinhardt, Gaea. "On Teaching." In *Advances in Instructional Psychology,* vol. 4, edited by Robert Glaser, pp. 1–54. Hillsdale, N.J.: Lawrence Erlbaum Associates, 1993.

———. "What Research on Learning Tells Us about Teaching." *Educational Leadership* 49 (April 1992): 20–25.

Leinhardt, Gaea, and Michael D. Steele. "Seeing the Complexity of Standing to the Side: Instructional Dialogues." *Cognition and Instruction* 23, no. 1 (2005): 87–163.

Linchevski, Liora, and Bilha Kutscher. "Tell Me with Whom You're Learning, and I'll Tell You How Much You've Learned: Mixed-Ability versus Same-Ability Grouping in Mathematics." *Journal for Research in Mathematics Education* 29 (November 1998): 533–54. Reprinted in *Lessons Learned from Research*, edited by Judith Sowder and Bonnie Schappelle, pp. 47–62. Reston, Va.: National Council of Teachers of Mathematics, 2002.

Livingston, Carol, and Hilda Borko. "High School Mathematics Review Lessons: Expert-Novice Distinctions." *Journal for Research in Mathematics Education* 21 (November 1990): 372–87.

Loucks-Horsley, Susan, Nancy Love, Katherine E. Stiles, Susan Mundry, and Peter W. Hewson. *Designing Professional Development for Teachers of Science and Mathematics,* 2nd ed. Thousand Oaks, Calif.: Corwin Press, 2003.

Loveless, Tom. "The Tracking and Ability Grouping Debate." July 1, 1998. http://www.edexcellence.net/foundation/publication/publication.cfm?id=127#802 (accessed June 28, 2006).

Ma, Liping. *Knowing and Teaching Elementary Mathematics: Teachers' Understanding of Fundamental Mathematics in China and the United States*. Mahwah, N.J.: Lawrence Erlbaum Associates, 1999.

Mathematical Sciences Education Board. *Knowing and Learning Mathematics for Teaching.* Washington, D.C.: National Academy Press, 2001. Also available online at http://www.nap.edu/books/0309072522/html/ (accessed June 21, 2006).

McGaffrey, Daniel F., Laura S. Hamilton, Brian M. Stecher, Stephen P. Klein, Delia Bugliari, and Abby Robyn. "Interactions among Instructional Practices, Curriculum, and Student Achievement: The Case of Standards-Based High School Mathematics." *Journal for Research in Mathematics Education* 32 (November 2001): 493–517.

Mewborn, Denise S., ed. *Teachers Engaged in Research.* 4 Vols. Greenwich, Conn.: Information Age Publishing and National Council of Teachers of Mathematics, 2006.

Mirra, Amy. *Administrator's Guide: How to Support and Improve Mathematics Education in Your School.* Reston, Va.: National Council of Teachers of Mathematics and Association for Supervision and Curriculum Development, 2003.

Mullis, Ina V. S., Michael O. Martin, Eugenio J. Gonzalez, and Steven J. Chrostowski. *Findings from IEA's Trends in International Mathematics and Science Study at the Fourth and Eighth Grades.* Chestnut Hill, Mass.: TIMSS and PIRLS International Study Center, Boston College, 2004. Also available online at http://timss.bc.edu/timss2003i/mathD.html (accessed June 28, 2006).

Mullis, Ina V. S., Michael O. Martin, Eugenio J. Gonzalez, Kelvin D. Gregory, Robert A. Garden, Kathleen M. O'Connor, Steven J. Chrostowski, and Teresa A. Smith. *TIMSS 1999 International Mathematics Report: Findings from IEA's Repeat of the Third International Mathematics and Science Study at the Eighth Grade.* Chestnut Hill, Mass.: International Study Center, Lynch School of Education, Boston College, 2000. Also available online at http://isc.bc.edu/timss1999i/math_achievement_report.html (accessed June 28, 2006).

Mullis, Ina V. S., Michael O. Martin, Eugenio Gonzalez, Kathleen M. O'Connor, Steven J. Chrostowski, Kelvin D. Gregory, and

Robert A. Garden. *Mathematics Benchmarking Report, TIMSS 1999—Eighth Grade: Achievement for U. S. States and Districts in an International Context.* Boston, Mass.: International Association for the Evaluation of Educational Achievement, 2001. Also available online at http://timss.bc.edu/isc/publications.html (accessed June 28, 2006).

National Commission on Mathematics and Science Teaching for the 21st Century. *Before It's Too Late: A Report to the Nation from the National Commission on Mathematics and Science Teaching for the 21st Century.* Washington, D.C.: U.S. Department of Education, 2000. Also available online at http://www.ed.gov/inits/Math/glenn/index.html (accessed June 28, 2006).

National Council for Accreditation of Teacher Education. *Professional Standards Accreditation of Schools, Colleges, and Departments of Education: 2006 Edition.* Washington, D.C.: National Council for Accreditation of Teacher Education, 2006. Also available online at http://www.ncate.org/institutions/standards.asp?ch=9 (accessed June 28, 2006).

National Council of Teachers of Mathematics (NCTM). *Curriculum Focal Points for Prekindergarten through Grade 8: A Quest for Coherence.* Reston, Va.: NCTM, 2006.

———. "Closing the Achievement Gap." Position statement. April 2005. http://www.nctm.org/about/position_statements/position_achievementgap.htm (accessed June 28, 2006).

———. *Perspectives on Teaching Mathematics: 2004 Yearbook,* edited by Rheta N. Rubenstein. Reston, Va.: NCTM, 2004.

———. *Principles and Standards for School Mathematics.* Reston, Va.: NCTM, 2000.

———. *Multicultural and Gender Equity in the Mathematics Classroom: The Gift of Diversity: 1997 Yearbook,* edited by Janet Trentacosta. Reston, Va.: NCTM, 1997.

———. *Assessment Standards for School Mathematics.* Reston, Va.: NCTM, 1995.

———. *Professional Standards for Teaching Mathematics*. Reston, Va.: NCTM, 1991.

———. *Curriculum and Evaluation Standards for School Mathematics*. Reston, Va.: NCTM, 1989.

———. *The Ideas of Algebra, K–12: 1988 Yearbook,* edited by Arthur F. Coxford. Reston, Va.: NCTM, 1988.

———. *An Agenda for Action: Recommendations for School Mathematics of the 1980s.* Reston, Va.: NCTM, 1980.

National Council of Teachers of Mathematics (NCTM) and National Association for the Education of Young Children (NAEYC). "Early Childhood Mathematics: Promoting Good Beginnings." Joint position statement. April 2002. http://www.naeyc.org/resources/position_statements/psmath.htm (accessed June 20, 2006).

National Research Council. *Helping Children Learn Mathematics*, edited by Jeremy Kilpatrick and Jane Swafford, Mathematics Learning Study Committee, Center for Education, Division of Behavioral and Social Sciences and Education. Washington, D.C.: National Academy Press, 2002. Also available online at http://www.nap.edu/catalog/10434.html (accessed June 20, 2006).

———. *Adding It Up: Helping Children Learn Mathematics*, edited by Jeremy Kilpatrick, Jane Swafford, and Bradford Findell, Mathematics Learning Study Committee, Center for Education, Division of Behavioral and Social Sciences and Education. Washington, D.C.: National Academy Press, 2001. Also available online at http://www.nap.edu/catalog/9822.html (accessed June 20, 2006).

———. *Educating Teachers of Science, Mathematics, and Technology: New Practices for the New Millennium.* Washington, D.C.: National Academy Press, 2000. Also available online at http://www.nap.edu/books/0309070333/html/ (accessed June 20, 2006).

Neuman, Mary, and Warren Simmons. "Leadership for Student Learning." *Phi Delta Kappan* 82 (September 2000): 9–12.

Organisation for Economic Co-operation and Development (OECD). "Where Immigrant Students Succeed—a Comparative Review of Performance and Engagement in PISA 2003." Paris, France: OECD, 2006. Also available online at http://www.pisa.oecd.org/document/44/0,2340,en_32252351_32236173_36599916_1_1_1_1,00.html (accessed June 22, 2006).

Phi Delta Kappa Educational Foundation. "Professional Book Study Program: How to Conduct a Book Study Group." 2003. http://www.pdkintl.org/bookstudy/home.html (accessed June 28, 2006).

Resnick, Lauren B. "Treating Mathematics as an Ill-Structured Discipline." In *The Teaching and Assessing of Mathematical Problem Solving,* edited by Randall I. Charles and Edward A. Silver, pp. 32–60. Hillsdale, N.J.: Lawrence Erlbaum Associates, 1988.

Reynolds, Anne. "What Is Competent Beginning Teaching? A Review of the Literature." *Review of Educational Research* 62 (Spring 1992): 1–35.

Reys, Robert, Barbara Reys, Richard Lapan, Gregory Holliday, and Deanna Wasman. "Assessing the Impact of Standards-Based Middle Grades Mathematics Curriculum Materials on Student Achievement." *Journal for Research in Mathematics Education* 34 (January 2003): 74–95.

Richardson, Joan. "Lesson Study: Teachers Learn How to Improve Instruction." *Tools for Schools* (February/March 2004). Also available online at http://www.nsdc.org/library/publications/tools/tools2-04rich.cfm (accessed June 28, 2006).

Riordan, Julie E., and Pendred E. Noyce. "The Impact of Two Standards-Based Mathematics Curricula on Student Achievement in Massachusetts." *Journal for Research in Mathematics Education* 32 (July 2001): 368–98.

Schifter, Deborah, Susan Jo Russell, and Virginia Bastable. "Teaching to the Big Ideas." In *The Diagnostic Teacher: Constructing New Approaches to Professional Development,* edited by Mildred Z. Solomon, pp. 22–47. New York: Teachers College Press, 1999.

Schoen, Harold L., Kristin J. Cebulla, Kelly F. Finn, and Cos Fi. "Teacher Variables That Relate to Student Achievement When Using a Standards-Based Curriculum." *Journal for Research in Mathematics Education* 34 (May 2003): 228–59.

Schoenfeld, Alan H. "Toward a Theory of Teaching-in-Context." *Issues in Education* 4, no. 1 (1998): 1–94.

Schoenfeld, Alan H., Jim Minstrell, and Emily van Zee. "The Detailed Analysis of an Established Teacher Carrying Out a Non-Traditional Lesson." *Journal of Mathematical Behavior* 18 (March 2000): 281–325.

Secada, Walter G. "Introduction." In *Changing the Faces of Mathematics: Multiculturalism and Gender Equity,* edited by Walter Secada, pp. 1–4. Reston, Va.: National Council of Teachers of Mathematics, 2000a.

Secada, Walter G., ed. *Changing the Faces of Mathematics: Multiculturalism and Gender Equity*. Reston, Va.: National Council of Teachers of Mathematics, 2000b.

Shulman, Lee S. "Theory, Practice, and the Education of Professionals." *Elementary School Journal* 98 (May 1998): 511–27.

———. "Knowledge and Teaching: Foundations of the New Reform." *Harvard Educational Review* 57 (February 1987): 1–22.

Slavin, Robert E. "Achievement Effects of Ability Grouping in Secondary Schools: A Best-Evidence Synthesis." *Review of Educational Research* 60 (October 1990): 471–99.

Smith, Margaret S. *Practice-Based Professional Development for Teachers of Mathematics*. Reston, Va.: National Council of Teachers of Mathematics, 2001.

Smith, Margaret S., and Mary K. Stein. "Selecting and Creating Mathematical Tasks: From Research to Practice." *Mathematics Teaching in the Middle School* 3 (February 1998): 344–50.

Steen, Lynn Arthur. "Back to the Future in Mathematics Education: Recent Reports Show That Little Has Changed in Mathematics Instruction." *Education Week* 23, no. 30 (April 7, 2004): 34, 36.

———. "Twenty Questions about Mathematical Reasoning." In *Developing Mathematical Reasoning in Grades K–12,* edited by Lee Stiff, pp. 270–85. Reston, Va.: National Council of Teachers of Mathematics, 1999.

Stein, Mary K. "Study of the Influence and Impact of *Principles and Standards for School Mathematics:* The Role of Schools and Districts in Teacher Capacity Building." In *Proceedings of the NCTM Research Catalyst Conference,* edited by Frank K. Lester Jr. and Joan Ferrini-Mundy, pp. 83–98. Reston, Va.: National Council of Teachers of Mathematics, 2004.

Stein, Mary K., Barbara W. Grover, and Marjorie Henningsen. "Building Student Capacity for Mathematical Thinking and Reasoning: An Analysis of Mathematical Tasks Used in Reform Classrooms." *American Educational Research Journal* 33 (Summer 1996): 455–88.

Stein, Mary K., and Margaret S. Smith. "Mathematical Tasks as a Framework for Reflection." *Mathematics Teaching in the Middle School* 3 (January 1998): 268–75.

Stein, Mary K., Margaret S. Smith, Marjorie Henningsen, and Edward A. Silver. *Implementing Standards-Based Mathematics Instruction: A Casebook for Professional Development*. New York: Teachers College Press, 2000.

Stigler, James W., and James Hiebert. *The Teaching Gap: Best Ideas from the World's Teachers for Improving Education in the Classroom*. New York: The Free Press, 1999.

Taylor, P. Mark. "Encouraging Professional Growth and Mathematics Reform through Collegial Interaction." In *Perspectives on the Teaching of Mathematics: Sixty-Sixth Yearbook,* edited by Rheta N. Rubenstein, pp. 219–28. Reston, Va.: National Council of Teachers of Mathematics, 2004.

Teacher Education Accreditation Council (TEAC). *Guide to Accreditation*. Washington, D.C.: TEAC, 2005. Also available online at www.teac.org/literature/index.asp (accessed June 23, 2006).

Tomlinson, Carol Ann. "Traveling the Road to Differentiation in Staff Development." *Journal of Staff Development* 26 (Fall 2005): 8–12.

Trentacosta, Janet, and Margaret Kenney (eds.). *Multicultural and Gender Equity in the Mathematics Classroom: The Gift of Diversity.* Reston, Va.: National Council of Teachers Mathematics, 1997.

Uecker, Jeffrey, and Douglas Cardell. "Decked Classes: Structuring the Mathematics Program for Radical Heterogeneity." In *Changing the Faces of Mathematics: Multiculturalism and Gender Equity,* edited by Walter G. Secada, pp. 115–23. Reston, Va.: National Council of Teachers of Mathematics, 2000.

United Federation of Teachers Teacher Center. "Four Key Practices." *Inside Professional Development* 1, no. 1 (Fall 2002). www.ufttc.org/publications/insidepd/insidepd6.htm (accessed June 23, 2006).

U.S. Census Bureau. "Population Profile of the United States: 1999." Current Population Reports, Series P23-205. Washington, D.C.: U.S. Government Printing Office, 2001. Also available online at http://www.census.gov/prod/www/abs/popula.html (accessed June 22, 2006).

von Glasersfeld, Ernst. *Radical Constructivism: A Way of Knowing and Learning*. London: The Falmer Press, 1995.

Wood, Terry, Paul Cobb, and Erna Yackel. "Change in Teaching Mathematics: A Case Study." *American Educational Research Journal* 28 (Fall 1991): 587–616.

York-Barr, Jennifer, William A. Sommers, Gail S. Ghere, and Jo Montie. *Reflective Practice to Improve Schools: An Action Guide for Educators*. Thousand Oaks, Calif.: Corwin Press, 2001.

Additional NCTM Resources for the Professional Development of Teachers of Mathematics

Readers of *Mathematics Teaching Today* can find many valuable resources for the professional development of teachers of mathematics among the publications available from NCTM, including the following:

- ***Perspectives on the Teaching of Mathematics (with Professional Development Guidebook), 66th Yearbook,*** edited by Rheta N. Rubenstein (Reston, Va.: National Council of Teachers of Mathematics, 2004). Organized around three primary aspects of teaching—foundations for teaching, the enactment of teaching, and the support of teaching—this book is designed to further teachers' growth and development in implementing effective mathematics instruction.

- ***Practice-Based Professional Development for Teachers of Mathematics,*** edited by Margaret Schwan Smith (Reston, Va.: National Council of Teachers of Mathematics, 2001). This book describes how to design, conduct, and evaluate professional education experiences for teachers. It explores a specific type of professional development opportunity that connects the ongoing professional development of teachers with the actual work of teaching.

- ***Designing Professional Development for Teachers of Science and Mathematics, second edition,*** by Susan Loucks-Horsley, Nancy Love, Katherine E. Stiles, Susan Mundry, and Peter W. Hewson, (Corwin Press, Inc., 2003). This book discusses the practices and issues of professional development for mathematics and science educators, examines the challenges ahead, and offers glimpses of the educational systems of the future. It espouses the principles of effective professional development and describes ways to create teacher learning programs based on those principles.

Please consult www.nctm.org/catalog for the
availability of these titles and for a plethora of resources
for teachers of mathematics at all grade levels.

❖

For the most up-to-date listing of NCTM resources on topics of interest
to mathematics educators, as well as on membership benefits, conferences,
and workshops, visit the NCTM Web site at www.nctm.org.